MODERN WARFARE

SOVIET COLD WAR WEAPONRY

Tanks and Armoured Fighting Vehicles

This Iraqi BMD-1 was destroyed in 2003 during Operation Iraqi Freedom. Iraq was only supplied a small number of these vehicles.

MODERN WARFARE

SOVIET COLD WAR WEAPONRY

Tanks and Armoured Fighting Vehicles

Anthony Tucker-Jones

Pen & Sword
MILITARY

First published in Great Britain in 2015 by
PEN & SWORD MILITARY
an imprint of
Pen & Sword Books Ltd,
47 Church Street,
Barnsley,
South Yorkshire
S70 2AS

ISBN 978 178303 296 9

Typeset by CHIC GRAPHICS

Printed and bound by Imago, China

Pen & Sword Books Ltd incorporates the imprints of Pen & Sword
Archaeology, Atlas, Aviation, Battleground, Discovery, Family History, History,
Maritime, Military, Naval, Politics, Railways, Select, Social History, Transport,
True Crime, Claymore Press, Frontline Books, Leo Cooper, Praetorian Press,
Remember When, Seaforth Publishing and Wharncliffe.

For a complete list of Pen & Sword titles please contact
Pen & Sword Books Limited
47 Church Street, Barnsley, South Yorkshire, S70 2AS, England
E-mail: enquiries@pen-and-sword.co.uk
Website: www.pen-and-sword.co.uk

Contents

Preface: Modern Warfare Series

Pen & Sword's Modern Warfare series is designed to provide a visual account of the defining conflicts of the late twentieth and early twenty-first centuries. These include Operations Desert Storm, Iraqi Freedom and Enduring Freedom. A key characteristic of all three, fought by coalitions, is what has been dubbed 'shock and awe', whereby superior technology, air supremacy and overwhelming firepower ensured complete freedom of manoeuvre on the ground in the face of a numerically stronger enemy. The focus of this series is to explain how military and political goals were achieved so swiftly and decisively.

Another aspect of modern warfare is that it is conducted in the full glare of the international media. This is a trend that first started during the Vietnam War and to this day every aspect of a conflict is visually recorded and scrutinised. Such visual reporting often shapes public perceptions of conflict to a far greater extent than politicians or indeed generals.

All the photos in this book, unless otherwise credited, were issued by the US Department of Defense at the time of the conflict. The author and the publishers are grateful for the work of the various forces combat photographers.

Introduction:
The Cold War

It is difficult today to remember that at the height of the Cold War the possibility of Communist hordes pouring across Central Europe was a very real threat. For four decades Europe stood on the brink of the Third World War, thanks to the heavily-armed standoff between the North Atlantic Treaty Organisation (NATO) and the Warsaw Pact. Thankfully it was the war that never was. The Cold War became a historical footnote, sandwiched between the Second World War and the conflicts of the early twenty-first century. It is one of those intriguing 'what ifs?' of history.

Washington never allowed its NATO allies to forget the extent of the Soviet threat. Annually throughout the 1980s the US Department of Defense published its *Soviet Military Power*, which catalogued Moscow's strategic aspirations and its latest military developments. Anyone reading it was left feeling that war was imminent and woe betide NATO if it was not ready.

By the mid-1980s the Cold War was at its height, with a conventional and nuclear standoff across Europe divided by the Iron Curtain. As part of its forward defence Moscow deployed armies in Eastern Europe with the Group of Soviet Forces in Germany, the Northern Group in Poland, the Southern Group in Hungary and the Central Group in Czechoslovakia. This not only guarded against NATO but also ensured none of the other Warsaw Pact members could defect. These forces were used to stop a repeat of the anti-Soviet uprising in East Germany of 1953, the Hungarian Revolt of 1956 and the Prague Spring of 1968. The following year the Soviet armed forces were involved in a Sino-Soviet border conflict and in 1979 became embroiled in a ten-year struggle in Afghanistan.

After the Second World War with tensions mounting between the Western allies and the Soviets, Berlin remained divided between the American, British and French sectors that made up West Berlin and the Soviet sector that occupied the east. This resulted in the Soviet blockade of West Berlin from June 1948 to May 1949. In response the Allies organised the Berlin airlift and war in Europe was only narrowly avoided. However, the Cold War went hot around the world, most notably in 1950 with the conflict in Korea.

The Warsaw Pact of 1955 brought together eight communist states in Central and Eastern Europe. Moscow argued the pact was a defensive move in light of West Germany being allowed into NATO. The reality was that it bound Eastern Europe's militaries to the Soviet armed forces. The Soviet Union was divided into military districts, with the key ones being the Baltic, Leningrad, Moscow and Kiev. By this stage the Soviet ground forces consisted of over 200 divisions, down from 500 at the end of the Second World War.

Not only did the Soviets have the numbers, they also had a vast array of weaponry. If there was one thing the Soviet Union was particularly good at it was building tanks. Since the mid-1950s Soviet-designed tanks dominated every single conflict right up until the 1991 Gulf War. Two designs in particular proved to be Moscow's most reliable workhorses – these are the T-54 and T-62 main battle tanks (MBTs). They are direct descendants of the Soviet Union's war-winning T-34 and Joseph Stalin tanks. They drew on the key characteristics of being easy to mass-produce, extremely robust and easy to use. As a result they were ideal for the less-well educated armies of the developing world. Having been inside a Czech-built T-54 I can testify that they are certainly no-frills tanks. The finish is not good and there are no creature comforts – clearly a legacy from the Spartan conditions inside the T-34. Nonetheless, they did the job that was required of them.

The scale of Soviet armour manufacturing at its height was immense. The tank plant at Nizhniy Tagil was supported by at least three other key tank factories at Kharkov, Omsk and Chelyabinsk, while other armoured fighting vehicles (AFVs) were manufactured at seven different sites. In the 1980s the Soviets were producing approximately 9,000 tanks, self-propelled guns and armoured personnel carriers/infantry fighting vehicles (APCs/IFVs) a year. The Soviet Union's East European Warsaw Pact allies managed another 2,500.

Moscow sent almost 8,000 tanks and self-propelled guns and over 14,000 APCs/IFVs to the developing world during that decade alone. In effect they exported two and a half years' worth of production. The Soviets' ability to manufacture such vast numbers of tanks meant that on at least two occasions they were able to save Arab armies from complete disaster at the hands of the Israelis.

By the 1980s Moscow had a staggering 52,600 tanks and 59,000 APCs in its active inventory, with another 10,000 tanks and APCs in storage. After the Warsaw Pact force-reduction talks in Eastern Europe, in 1990 Moscow agreed to withdraw 10,000 tanks and destroy half of these without batting an eyelid. Warsaw Pact members also agreed to cut tank numbers by almost 3,000. At the same time the Soviets began to field newer tanks such as the T-64B, T-72M1 and the T-80, while retiring older-model T-54/55s and T-62s. They also improved their IFV forces by fielding large numbers of the tracked BMP-2 as well as improving the earlier BMP-1. The net result was a huge surplus of wheeled AFVs available to the developing world.

The British Army of the Rhine (BAOR) was once part of the bulwark that helped protect Western Europe from the threat posed by the Soviet groups of forces stationed across Eastern Europe and their Warsaw Pact allies. At the height of the Cold War BAOR, serving with NATO's northern army group, represented the largest concentration of ground forces in the British Army. It consisted of the isolated Berlin Independent Brigade and the 1st British Corps in West Germany. HQ BAOR was based at Rheindahlen while HQ 1 (BR) Corps was at Bielefeld, commanding three divisions.

The fate of the American, British and French garrisons in West Berlin had the Cold War gone hot would have been certain. It is likely that the Warsaw Pact would have first cut them off and then overwhelmed them. But this never came to pass, however; West Germany and East Germany along with the two halves of Berlin were reunited on 3 October 1990. The following year the Soviet Union collapsed and the Cold War came to an end

While the Cold War resulted in an armed standoff either side of the Iron Curtain, Moscow actively supported the spread of Communism, elsewhere most notably in Korea and Vietnam. Tanks with one previous owner, no strings attached (except when that previous owner happened to be the Soviet Union, there were always strings attached). The fact that the tank was ancient, would not meet your operational requirements and leave you heavily indebted to Moscow did little to deter many developing countries desperate for huge quantities of weapons. From the Horn of Africa to Central America, the Soviet T-55 and T-62 MBTs became as ubiquitous as the Kalashnikov AK-47 assault rifle.

Although the two Superpowers were cautious about coming into direct confrontation, this did not prevent indirect meddling elsewhere in the world. On the periphery, the Cold War became very hot and on a number of occasions almost sparked war in Europe. Time after time Moscow was able to make good its allies' massive losses. The Soviets conducted a substantial re-supply of Syria in 1982–3 following their military losses in Lebanon. Major re-supply also took place in 1977–9 in support of Ethiopia in its clash with Somalia and during the Arab-Israeli Wars of 1967 and 1973. Prior to that they conducted airlift operations in 1967–8 in support of a republican faction in North Yemen.

At the height of the Cold War the Soviet Union exported billions of dollars worth of arms to numerous developing countries. Intelligence analysts watched with a mixture of alarm and awe as cargo ship after cargo ship sailed from Nikolayev in Ukraine stacked to the gunnels to ports such as Assab in Ethiopia, Luanda in Angola, Tartus in Syria and Tripoli in Libya. Much of this equipment came from strategic reserves and was very old or had been superseded by newer models, as in the case of the T-55 and T-62 MBTs, which were all but obsolete by then. Soviet armoured vehicle exports also included the 4x4 wheeled BTR-60 APC and the tracked BMP-1 IFV.

In many cases Soviet weapon shipments were funded through generous loans, barter-deals or simply gifted, and Moscow's arms industries rarely saw a penny in return. The net result was that during the Cold War Moscow fuelled a series of long-running regional conflicts that lasted for decades. Ultimately the West was to spend the Soviet Union into oblivion, but the legacy of the Cold War was one of global misery.

Chapter One

T-54/55 Main Battle Tank

On 23 October 1956, elements of two motor rifle divisions from the Soviet garrison in Hungary entered Budapest to forestall Hungarian attempts to throw off Moscow's domination. Two further divisions moved across the Hungarian-Romanian border to support them. These forces were equipped with the T-34 tank, victor of the Second World War. Their crews came in for a nasty surprise as the T-34s proved vulnerable to Molotov cocktails on the streets of Budapest and, in the face of unexpectedly strong resistance, the Soviet troops were forced to withdraw.

However, Moscow was not deterred in the least. On 4 November 1956 a fresh phase of fighting started when more Soviet troops were committed with the much newer T-54 MBT. The latter was not so easy to disable and the Hungarians were overwhelmed in ten days. The West, distracted by the Suez Crisis, failed to notice the baptism of fire of the Soviet Army's brand new tank, which was to become an icon of the Cold War.

From the Horn of Africa to Southeast Asia, the ubiquitous T-54/55 and T-62 tanks have been central to numerous regional conflicts fought since the Second World War. For many years they formed the armoured backbone of the Warsaw Pact armies allied to Moscow, and were also dispatched around the world to numerous Soviet client states. The T-54/55 has seen service with well over eighty countries while the T-62 ended up in service with around twenty. No other tanks have ever enjoyed such global success.

T-54/55s were delivered to countries such as Afghanistan, Angola, Egypt, Ethiopia, India, Iraq, Libya, Mozambique, Syria, Vietnam and Yemen, all of which were involved in numerous bloody wars. In particular, Afghanistan and Iraq became graveyards for thousands of abandoned and rusting T-54s and T-62s. The Israelis ended up with so many captured from the Arab armies that they put them into service themselves.

The Soviet Union's strategic reserves and production capabilities were such that they could swiftly re-supply its allies at very short notice. This proved vital during the Arab-Israeli Wars when the Arabs tended to lose vast numbers of tanks to superior

During the Cold War a considerable proportion of Moscow's tank fleet consisted of the T-54/T-55 MBT. Around 50,000 were built between 1954 and 1981 and many were issued to Soviet motor rifle divisions.

T-55s of the Ugandan Army on parade during the 1970s. At this time the Ugandans also fielded limited numbers of T-34 and PT-76 tanks.

T-54/55 tanks captured by the Israelis during the Arab-Israeli wars. Many of these were refurbished and put into service with the Israeli army.

Afghan Mujahideen with a captured T-54. Both the Soviet and Afghan armies used this tank type during the Soviet-Afghan War.

An Israeli-supplied upgraded T-55 known as the Tiran-5 abandoned by the South Lebanese Army.

Ethiopian T-55s outside the presidential palace in Addis Ababa following the fall of the government in 1991. Ethiopia was blighted by civil war throughout the 1970s and 1980s.

Although the T-54 went into production in the late 1940s, it was dubbed the T-54. It is readily identifiable from the subsequent improved T-55 by the frying pan-like ventilator dome on the left-hand side of the turret – clearly visible on these Cambodian Army T-54s.

Israeli tactics and gunnery. On a number of occasions Moscow was able to stave off their defeat by conducting a massive re-supply operation.

Originally the T-54 and the T-62 were designed to overwhelm the forces of NATO on the central plain of Germany if the Warsaw Pact armies ever stormed through the Fulda Gap. As a result they were squat, offering the lowest profile possible, the reasoning being they would have to close with NATO's ground forces as swiftly as possible. In contrast, NATO's tanks were designed to keep the enemy at arm's-length, so presented a much higher silhouette to give the tank gunners greater visibility and range. The T-54 and T-62's low silhouette and therefore much reduced gun depression/elevation was to prove a distinct disadvantage when fighting amongst the sand dunes of the Middle East during the Arab-Israeli Wars.

In the closing years of the Second World War the Soviet Union designed a new medium tank called the T-44 that sought to improve upon the highly battle-proven T-34/76 and T-34/85 which carried the Red Army to Berlin. The T-44 only appeared in very limited numbers between 1945 and 1949, seeing service at the end of the Second World War and then during the Hungarian Uprising in 1956.

It was followed by the T-54, the first prototype appearing in 1946 with production commencing the following year in Kharkov. It had a very distinctive mushroom-shaped turret that drew on that of the Joseph Stalin heavy tank, which provided excellent shot-deflection surfaces. The all-welded T-54 hull consisted of three compartments, driver's at the front, fighting compartment in the middle and engine/transmission in the rear.

The round turret was a one-piece casting with the top comprising two D-shaped pieces of armour welded together down the middle. The commander sat on the left of the turret with the gunner on the same side but in a more forward position. The commander's cupola could be traversed through 360 degrees, with a single-piece hatch that opened forward with a single periscope on each side. A TPK-1 sight with a single periscope either side was mounted in the forward part of the top of the cupola. The loader sat on the right of the turret and had a periscope and a single hatch that opened to the rear.

An early photograph of the T-54-2 on manoeuvres. It went into production in 1949 and was followed by the T-54A and T-54B. At the time of the Soviet Union's collapse Moscow was believed to still have almost 20,000 T-54/55s in its inventory. Many have since been scrapped or sold off.

A Soviet AFV crewman in the standard tanker's padded helmet – the gold CA (for Soviet Army) identifies that he is not a tanker, otherwise he would have a tank symbol.

The driver sat at the front of the tank to the left and had a single-piece hatch that swung to the left. There were two periscopes forward of this hatch, one of which could be replaced by an infra-red periscope which was used in conjunction with the infra-red searchlight mounted on the right side of the glacis plate. To the right of the driver was an ammunition stowage space, batteries and a small fuel tank.

The T-54 engine was mounted in the rear of the hull and the tank used an electrical start-up system with a compressed-air system as a back-up in cold weather. In contrast the subsequent T-55 used a compressed-air engine starter system, with an electrical back-up. This was because, unlike the T-54, the T-55 had an AK-150 air compressor to refill the air-pressure cylinders.

The T-55 appeared in 1958 and was essentially the T-54 with a new turret without the distinctive rooftop ventilator dome. It also had a new stabiliser, the ammunition load increased to forty-three rounds (up from thirty-four), new running gear and a more powerful V-55 diesel engine that gave slightly greater horsepower, though the speed of 50km/h remained the same. The T-54/T-55 series had a torsion-bar suspension that consisted of five road wheels with a very distinctive gap between the first and second road wheel. The drive sprocket was at the rear and the idler at the front. Neither the T-54 nor the T-55 had track return rollers.

Similar to the T-34 tank, the T-54/55's all-steel tracks had steel pins that were not held at the outer edge and therefore travelled towards the hull. A raised piece of metal welded to the hull just forward of the sprocket drove the track pins back in every time they passed.

A number of bridgelayers employed the T-55 chassis, including the Soviet MTU-20, the Czechoslovak MT-55 seen here and the jointly-developed East German and Polish BLG-60. Such equipment formed an integral part of Soviet armoured divisions.

The T-54 and the upgraded T-55 had frontal armour of just under 100mm and a range of around 500km. The T-54 was produced in at least eight different variants, while the T-55 had up to a dozen variants, though outwardly the differences in appearance were minimal. Engineering variants included armoured recovery vehicles, bridgelayers, dozers and mineclearers. Space does not permit a detailed listing of all the variants and their specifications.

Both were armed with a 100mm gun, with the flume extractor very near the muzzle. It fired ordinary armour-piercing (AP) shells, which give limited penetration at long range. AP is a solid full-calibre steel shot, which dissipates much of its energy before reaching the target. During the Arab-Israeli Wars the Israelis had armour-piercing discarding sabot (APDS), and high explosive anti-tank (HEAT) ammunition that provided kills at far greater ranges. The sabot packing around the shell is stripped off by air resistance to reveal an arrow of metal that offers far greater penetration. HEAT, by contrast, uses a jet of molten copper to penetrate through to the interior of a tank with predictable results.

The T-54 gunner had to estimate the range with visual adjustment, dubbed

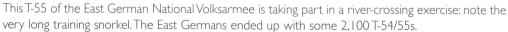

This T-55 of the East German National Volksarmee is taking part in a river-crossing exercise: note the very long training snorkel. The East Germans ended up with some 2,100 T-54/55s.

This Polish T-55 is also conducting a river crossing; by the late 1980s Poland had amassed 2,700 T-54/55s. Visually, the T-55 is essentially the same as the T-54: for recognition purposes though the two tanks were often grouped together generically as the T-54/55.

'Eyeball Mark I'. In contrast, American-built M48 and M60 tanks had accurate optical-prism rangefinding systems that allowed zeroing in on targets within seconds, while British-built Centurions used machine-gun tracer bullets to correct the main gun targeting.

The T-54/55 could ford rivers through the use of a snorkel. Two types were available, a thin one for operational use and a thick one for training. They took up to 30 minutes to fit and were blown off once the far riverbank was reached. The combat snorkel was mounted over the loader's periscope and when not fitted was stored disassembled at the rear of the hull or the turret.

While Moscow never released any official figures for T-54/55 production, it has been estimated that the Soviet Union alone built about 50,000, while those built in China, Czechoslovakia and Poland bring total numbers to around 72,000. This greatly

Allies on exercise – the man on the right is a Soviet tanker identifiable by his black overalls and yellow tank patch just visible embroidered on the right breast. The non-Soviet Warsaw Pact allies opted not to deploy the newer T-62 when it became available and instead relied on enormous fleets of T-54/55s seen here. When they did upgrade their tank divisions they selected the T-72.

This is a Czechoslovak-built T-54, which were said to be the best engineered. However, in this example the radio, optics, gauges and most of the electrics are Soviet. Note the gap between the first and second road wheel – on the T-62 there are gaps between the third and fourth and the fourth and fifth wheels.

exceeds the number of T-34s, even allowing for post-war T-34 construction by Czechoslovakia and Poland, of about 60,000.

Remarkably, the T-55 managed to outlast its successor the T-62. Production of the T-55 is thought to have run at the Omsk tank factory until 1981, long after the T-62 had gone out of production. Poland produced the T-54 from 1956 to 1964 and the T-55 from 1964 onward. The East German Army favoured the T-54 over the newer T-62 and declined to accept the latter when it came into service. Instead they waited until the T-72 had been produced before they upgraded.

An interesting example of a Czech T-54 is in the Cobbaton Combat Collection. Although Czech-built, many of the fittings including the radio, optics, gauges and most of the electrics, are Russian. It is still a T-54 specification, having never been uprated to T-55 standard. While the controls are heavy, it is a good tank to drive and it is said that the Czech versions are better-engineered than those built by the Russians.

In the early 1950s Moscow supplied China with a number of T-54s, the Chinese subsequently building a version themselves as the Type 59. Later models were fitted with a fume extractor similar to the T-54A. Subsequent upgrades resulted in the Type 59-I and Type 59-II, the latter being armed with a 105mm rifled gun. A further development of the Chinese Type 59 was the Type 69 that first appeared in public in 1982. Large numbers of both tank types were cynically exported to Tehran and Baghdad during the Iran-Iraq War in the 1980s. Pakistan also proved to be a major customer for both models. The differences in appearance between the Type 59 and Type 69 are minimal. Drawing on these designs the Chinese went on to produce the Type 79, 80, 85 and 90 tanks.

Romania produced the TR-85, which is very similar in appearance to the T-55 series. Locally-built Romanian T-55s were first seen in 1977 and were designated the TR-77 by the West. The key difference with the later TR-85 is that it has six road wheels while the T-55 has five. Nor does it have the common exhaust outlet on the left-hand side that is a standard feature of the T-54/T-55. Subsequent Romanian versions of the T-54/55 included the TR-580 and the TN-800, though it is unclear if these went into series production.

An Iraqi T-54/55 or Type 59 photographed during the Iran-Iraq War of the 1980s. Later, Iraq's Soviet and Chinese-supplied tanks proved to be all but useless in the face of the Coalition's overwhelming firepower during the first and second Gulf Wars.

A column of Polish Army T-55As on the streets following the declaration of martial law in Poland in 1981. The Polish government managed to head off Soviet intervention by taking matters into their own hands.

A T-55 supplied to the Somali National Army in the early 1980s. The Somali Democratic Republic also fielded the T-34.

In Europe the T-54 drew first blood on the streets of Hungary in 1956 when Moscow crushed the Hungarian Uprising. It first went into action in the Middle East with the Egyptian and Syrian armies in 1967, whereas the T-62 did not see combat until 1973. On the Indian subcontinent the T-54 and T-55 first saw combat in 1971 during the Indo-Pakistan War, serving with the armies of both sides. In Africa the Somalis used the T-54 against the Ethiopians during the 1977 Ogaden War and it subsequently saw action in Angola and Mozambique.

In the event of war between NATO and the Warsaw Pact in Central Europe the T-54/55s would have been of questionable value, despite their vast numbers. Their thin armour and 100mm gun would have proved little threat to such tanks as the British Chieftain, the German Leopard and the American Abrams. Such considerations did not stop the T-54/55 becoming one of the most ubiquitous tanks of all time. Whether NATO could have knocked out the hordes of T-54/55s and other Soviet tanks fast enough before they were overrun remains open to debate.

Civilians pose on a Libyan T-55 after it was captured by rebels in Benghazi in 2011. The Libyan Army had 500 T-55s in service and over a thousand T-54/55s in store.

Chapter Two

T-62 & T-64 Main Battle Tanks

T-62 MBT

The T-62 was designed at Nizhnyi Tagil, based on the T-55, and incorporated a number of its components, but had a longer and wider hull and a new turret. While the engine and transmission from the T-55 were retained, a larger-diameter fan improved the cooling system. The suspension was the same as the T-55 but the mounts were reconfigured to allow for a longer hull. This meant that the spacing for the road wheels on the T-62 was different to that of the T-55, with distinctive gaps between the third and fourth and fourth and fifth road wheels. Likewise a distinctive flume extractor was two-thirds up the barrel of the larger-calibre 115mm gun.

The T-62 was in production in the Soviet Union and Czechoslovakia from 1961 until 1975 and was widely exported. This photograph was first published in *Soviet Military Power* in 1984.

Although there were over a dozen different T-62 variants, including upgrades, the three key production models were the 1962, 1972 and 1975 versions. Although pre-production models of the T-62 were built in 1961, it was not seen publicly until 1965. Manufacture of the T-62 ran from 1961 to 1975 during which time around 20,000 were built, nowhere near the number of T-54/55s. The T-62 was supplied to over twenty countries. Czechoslovakia and North Korea also built it for domestic and export purposes, the Czechs constructing around 1,500 between 1973 and 1978.

As in the T-54/55 the cast turret sat in the middle of the tank with the commander and gunner on the left and the loader on the right. Both had a single-piece hatch that opened to the rear and could be locked vertically. The T-54 had twin hand rails on either side of the turret, while the T-62 had a single large hand rail either side that could be used by tank riders or for stowing personal equipment.

The commander's cupola had four periscopes, two mounted in the hatch cover

Soviet T-62 tank crews enjoy some refreshment courtesy of the locals. The easier-to-produce T-62 was deployed with the more numerous motor rifle divisions while the newer T-64 was issued to the tank divisions.

The T-62 shared many of the same design characteristics as the T-54/55 but had a longer and wider hull and a new turret. It is readily identified by the bore evacuator two-thirds of the way down the barrel. Though the Second World War tactic of tank *desants* or tank riders was obsolete, during the Cold War Soviet tanks continued to be produced with turret grab rails.

and two in the forward part of his cupola. The commander's sight, the TKN-3, was a day/night binocular periscope with an integral infra-red capability mounted in the forward part of his cupola. For daylight use it had a magnification of x5 and a 10-degree field of view and for darkness it had a magnification of x4.2 and an 8-degree field of view. The effective range when used with the OU-3GK infra-red searchlight was 400m. The handles of the sight were employed to rotate the commander's cupola and operate the searchlight and target-designation equipment.

A rear view of a T-62 that has been fitted with the spare 200-litre external fuel drums to extend its range. All vehicles had three external fuel cells on the right side for diesel fuel with a single tank on the left for auxiliary oil. Normally the driver would first use the drums, then the external fuel cells followed by the main internal fuel tank.

This T-62 is laying down smoke to conceal it from enemy gunners. This was done by spraying diesel oil into the hot exhaust manifold which created thick white smoke that came out of the left-hand exhaust ports. The smoke screen could be up to 400m long, lasting around four minutes, but this consumed 40 litres of fuel.

The T-62's main armament was a 115mm UT-5TS (2A20) smoothbore gun fitted with a bore evacuator, with a maximum rate of fire of four rounds a minute at a standstill. Mounted coaxially to the right of the main armament was a 7.62mm PKT machine gun that had a practical rate of fire of up to 250 rounds a minute, fed by a belt containing 250 rounds.

Once fired, the main gun automatically elevated for reloading, but the turret could not traverse while the gun was being loaded. An integral spent-shell ejection system activated by the recoil of the gun threw the empty cartridge case out of the turret through a trapdoor in the rear of the turret. The T-62 could carry forty rounds, with two ready rounds in the turret and the rest stored by the driver and in the rear of the fighting compartment. The gunner had a TSh2B-41u telescope with a rotating graticule for super elevation required for different types of ammunition and dual magnification, with x3.5 with an 18-degree field of view and x7 with a 9-degree field of view.

The T-62's main gun used high explosive (HE), HEAT or armour-piercing fin-stabilised discarding-sabot (APFSDS) ammunition which had a range of 4,000m. The HEAT round had a range of 3,700m while HE reached out to 4,800m. The T-62

An up-armoured T-62M of the 'Berlin' tank regiment, 5th Guards Motorised Rifle Division, leaving Afghanistan in 1987.

An Iraqi T-62 knocked out near the Iranian city of Khorramshr, Khuzestan province, during the Iran-Iraq War fought in the 1980s. At the start of the war the Iraqi Army had 2,500 T-54/55S and T-62s.

could match the guns of Israeli tanks but they were in short supply during the Yom Kippur War in 1973. The Egyptian Army only had about 100 T-62s as opposed to over 1,600 T-54/55s.

Basically, Soviet tanks were designed to present the minimum-sized target out in the open. For example, the T-54/55 has a much lower profile than the American M48 or British Centurion. The problem with this immediately became apparent; because of its low turret a Soviet tank gun has very limited depression. During the Arab-

Israeli Wars Israeli tanks had a distinct advantage in that they could depress their guns ten degrees below the horizontal, while the T-54 and T-62 could only go down four degrees. This, of course, offered distinct advantages when fighting in sand dunes or on rocky ridges of the Golan Heights, the Negev Desert and the Sinai, as a Soviet-built tank would have to expose itself to engage the enemy. This meant that in many instances Soviet tanks were unable to fight from a 'hull-down' position. As a result this presented real problems when fighting from defensive positions.

Despite such limitations, the T-62's main gun caused a nasty surprise in the West as it appeared at a time when most NATO armies had chosen to standardise on the 105mm calibre. The Soviet 115mm gun was not only larger, but was also a smoothbore, which was a major departure from the accepted rifled bore of the time. Nonetheless, the T-62 was far from perfect: it suffered from thin armour, vulnerable ammunition and fuel storage, a poor gearbox in early models, a tendency to shed its tracks and generally poor operating conditions for the crew.

T-64 MBT

The T-62's larger cousin, the T-64, appeared in the mid-1960s, but only about 8,000 were built and none were ever exported. The first prototype was finished in 1960 with the second three years later. The first production run was completed in 1966 with about 600 tanks that were all armed with the 115mm smoothbore gun. These suffered problems with the automatic loader, power pack (particularly the transmission) and the suspension. As the T-64 featured an automatic loading system, the crew could be reduced to three men, helping to keep the size and weight of the tank down.

Another innovation on the T-64 that was less successful was the suspension. All Soviet medium tanks from the T-34 on had used five road wheels without any return rollers, so why the change was made to six very small road wheels and four return rollers on the T-64 is not readily apparent, though it was known that the T-62 had a habit of losing its tracks. The T-64's design features appear to have failed, as the T-72 employed a completely different system, while modified T-62s were seen with the T-72-style suspension, not that of the T-64.

Confusingly the T-64 was very similar in appearance and layout to the T-72. The suspension consisted of six small dual road wheels (though these were notably smaller than the six used on the T-72) and four track return rollers (the T-72 only has three), with the idler at the front and the drive sprocket at the rear. The tracks were narrower than the T-72's and the turret was slightly different. The driver sat at the front in the centre, while the other two crew were located in the turret, with the commander on the right of the gun and the gunner to the left.

The follow-on T-64A sought to iron out the early design faults and included the

The T-64 appeared in the mid-1960s and was followed by the improved T-64A and T-64B. This photograph shows the difference between the T-64's road wheels and those of the T-72 in the background, which are much larger.

This T-64 is in East Germany in the 1980s serving with a tank division of the Western Group of Forces. Its road wheels look decidedly flimsy. Although around 10,000 were produced, T-64s only ever equipped the Soviet Army.

125mm 2A26M2 smoothbore gun fed by an automatic loader. This went into service in 1969 and was first seen publicly the following year during the Moscow Parade. The 125mm gun was stabilised in both elevation and traverse with the barrel fitted with a thermal sleeve and flume extractor. It could fire up to eight rounds a minute and had a sighted range out to 4,000m employing the day sight and 800m employing the night sight. The 2A26 gun had vertical ammunition stowage, while the T-72 and T-80 are armed with a 125mm 2A46 gun with a horizontal ammunition feed system. The gunner selected the type of ammunition he wished to fire by simply pushing a button. This was the separate loading type, in that the projectile is loaded first followed by the semi-combustible cartridge case; all that remains after firing is the stub base of the cartridge which is ejected. The 125mm ammunition is common to the T-64, T-72, T-80 and T-90 tanks.

The T-64B's 125mm gun could also fire the AT-8 'Songster' anti-tank guided weapon, which was kept in the automatic loader in two separate parts like standard APFSDS and HEAT-FS rounds and loaded using the automatic loader. The 12.7mm anti-aircraft machine gun on the T-64 could be aimed and fired from within the tank. In total around a dozen different T-64 variants were produced including command and up-armoured types.

This photograph shows how small the T-64's road wheels are. These and the suspension proved problematic and were not repeated in the T-72. The complete lack of T-64 exports shows how flawed it was and Moscow decided it was too expensive and complicated to supply to its client states.

Whereas the Soviets had gone for ease of mass-production with the T-54/55 and T-62, the latter's counterpart was much more advanced. As a result, the T-62 was assigned to the motor rifle divisions while the newer T-64 only served with the armoured divisions. Somewhat ironically, the T-64 entered production only slightly earlier than the T-72, which was intended to replace the T-54/55 and T-62. The T-64, while being a superior tank, suffered numerous teething problems that eventually consigned it to the scrap heap.

Although the T-64 served with the Soviet Groups of Forces stationed in the Warsaw Pact countries, it only ever saw combat against Chechen separatists. Only the Soviet Army employed it and with the break-up of the Soviet Union, the Russian Federation kept 4,000 of them while Ukraine ended up with 2,000. By 2013 Russia had scrapped all its T-64s, although Ukraine modernised some and kept them in service. None seem to have ever been exported.

The successors to the T-54/55 and T-62, the T-72, T-80 and T-90 have been produced in nothing like the same numbers nor have they been so widely exported. The T-54/55 and T-62 retain their status as Moscow's tried and tested workhorses.

Left and Above: Soviet tank crews on exercise with the T-72. This proved vastly more successful than the troubled T-64, which was very similar in appearance.

The T-72 was considered NATO's true nemesis and was widely exported.

Chapter Three

T-72 & T-80 Main Battle Tanks

T-72 MBT

Due to their shape, from a distance the T-64, T-72 and T-80 all look alike. The T-72 was a progressive development of the T-64 with improved suspension and a slightly different turret. The main difference was that the newer tank featured six large road wheels, whereas the T-64 had six rather small ones that were unlike those on any other Soviet tank. The T-72 has seen combat in well over twenty different conflicts and has served with over forty armies. Two key features of the T-72, and its counterpart the T-64, are its powerful 125mm gun (deployed at a time when most of its adversaries were sporting 105mm guns) and its relatively light weight. The 125mm gun

A very early intelligence photograph of the T-72, which like the T-54/55 and T-62 was to prove a great success. The spring-loaded skirt plates over the forward part of the tracks are locked in the travelling position.

The T-72 entered service with the Soviet and other Warsaw Pact armies in the 1970s. It was widely exported and was built under license in Czechoslovakia, India, Poland and the former Yugoslavia.

T-72s on parade commemorating the October Revolution in Moscow at the height of the Cold War. Starting with the T-72 (1973), there were in total some fifteen different variants designated by the Soviet Army.

fires APFSDS, HE or HEAT rounds and has an integrated fire-control system. This relieves both commander and gunner of some of their tasks as well as increasing the probability of a first-round hit. This of course is a crucial capability in a tank battle.

The T-72 came about in part as an attempt to develop a simpler MBT as an alternative to the complicated, expensive and somewhat disappointing T-64. In the 1960s a whole series of prototypes appeared, but the actual T-72 prototype was not completed until 1970. Essentially it drew on all its predecessors. It utilised the hull and turret layout of the T-64 as well as a similar drive train. The engine is an improved version of that used in the tried and tested T-62 and the cooling system is very similar to the one used in the T-55/T-62. All in all it was a successful hybrid that brought together the best elements of all the Soviet Army's previous tanks.

The driver is seated at the front of the hull and has a single-piece hatch that opens to the right, in front of which is a single wide-angle TVNE-4E observation periscope. As in the T-64, the other two crew members are seated in the turret, with the commander on the right and the gunner on the left. The commander's contra-rotating cupola has a single hatch that opens forward with two rear-facing TNPA vision blocks. In the front is a combined TKN-3 day/night sight with an OU-3 infra-

The T-72A seen here went into production in 1979 and was the fourth variant to enter service. This had the optical rangefinder replaced by the TPDK-1K laser rangefinder sight which greatly improved first-round hit probability. Along with other enhancements it had a significant increase in armour protection. The export version was known as the T-72M.

A Russian T-72 B3 sporting appliqué armour – the B series first appeared in the mid-1980s with the T-72BM appearing in 1992. The various variants included the T-72B, B1 and BK.

A T-72 on a low-loader transporter, the principal method of moving tanks over long distances.

red searchlight mounted over the top and either side of the combined sight is another TNP-160 periscope.

The gunner's hatch opens forward and has a circular opening for mounting the snorkel for deep fording. To the front of the gunner's hatch is a TNP-160 periscope, and a TNPA-65 vision block is installed in the hatch itself. In front and to the left of the gunner's hatch is a panoramic day/night sight, which is used in conjunction with the infra-red searchlight mounted to the left and in front of the sight. The gunner uses the TPD-2-49 day sight and the TPN-1-49-23 night sight.

The suspension on either side consists of six road wheels and three return rollers supporting just on the inside of the track, with the idler at the front and the drive sprocket at the rear. The standard production T-72s were fitted with four removable spring-loaded skirt plates on either side, fitted over the forward part of the track, which were unclipped in action and spring forward at an angle of 60 degrees from the side of the vehicle. These gave some protection against HEAT projectiles.

The T-72 is powered by a V-12 piston multi-fuel air-cooled engine that produces 740hp. It can run on three types of fuel, diesel, benzene and kerosene, with the

A monument to the T-72 in Nizhnyi Tagil, where it was developed and produced by the Ural tank plant. T-72s were also built at Chelybinsk and Kirov tank factories in Russia.

An Armenian T-72 memorial in Nagorno-Karabakh commemorating the war with Azerbaijan. Following the break-up of the Soviet Union, Armenia ended up with about 100 of these tanks, as did Azerbaijan.

driver being provided with a dial to set the engine for the type of fuel being carried. Later production models such as the T-72S were fitted with the much more powerful V-84 engine developing 840hp.

The main armament is the 125mm 2A46 smoothbore gun fitted with a light alloy thermal sleeve and a bore evacuator. It fires three types of separate-loading ammunition; APFSDS with a maximum range of 2,100m, high explosive anti-tank fin stabilised (HEAT-FS) with a direct fire range of 4,000m and high explosive fragmentation fin stabilised (HE-FRAG-FS) with an indirect fire range of 9,400m. The T-72 can carry a total of thirty-nine rounds of ammunition.

The tank went into production in 1972 and became fully operational the following year, but it was not seen in public until 1977. Around 20,000 T-72s were produced for both the home and export markets. Some countries were also permitted to set up T-72 production facilities – though these were usually to assemble Soviet-supplied knock-down kits rather than building the tank from scratch. Nonetheless, Moscow gave the T-72 design to a number of Warsaw Pact members so they could build the tank for their own armies and in some cases for export.

As a result, the T-72 has seen extensive service not only with the Soviets but also with numerous foreign armies, most recently in Syria where the Syrian Army fought to quell a widespread rebellion. In fact, Syrian T-72s were first blooded against the Israelis in Lebanon back in 1982. They were also deployed by Colonel Gaddafi's regime in 2011 in an unsuccessful attempt to crush the Libyan rebels. Iran and Iraq also used the T-72 against each other during the Iran-Iraq War in the 1980s.

In contrast to the Soviets, the British never quite got tank design right until the Centurion and the Challenger. The Centurion was a bit of an aberration – drawing on all their experience during the Second World War, British tank designers actually came up with a very good tank that proved to be a major export success. The Israelis in particular had great respect for the Centurion's tough capabilities and adapted it for their needs. Likewise, American tanks always tended to be slightly behind the design curve – the M4 Sherman epitomised this. While the Americans' range of Patton tanks were good, resulting in the M60, they still had their flaws. It was not until the advent of the Abrams that the Americans produced a truly dominant MBT.

It was the Soviets who got there first with the T-72. Its low silhouette in particular gives it a very sinister air. It was Soviet tank designers who pioneered the low-profile turret – normally the bigger the main gun the bigger the turret. The latter on any tank is the weak point as it is exposed and presents a ready target. The squat turrets of the American Abrams, British Challenger, French Leclerc and German Leopard are taken for granted, but it was the Soviets who pioneered it with the T-72.

During the 1970s and the 1980s, at the height of the Cold War, this tank was seen as NATO's nemesis in central Europe. The fear was that powerful Soviet armoured divisions would simply overwhelm the NATO armies and pour through the Fulda Gap. This never came to pass, but if it had Soviet T-72s would have been supported by Czech and Polish-built T-72s in the forefront of the fighting.

The T-72 was intended as an affordable way of replacing Moscow's existing workhorse tank fleet comprising the T-54/55 and T-62. It went into service with the Soviet and Warsaw Pact armies throughout the 1970s and was also widely exported. In total there were at least fifteen Soviet variants. Serving with the Soviet Army, the T-72 saw very limited service in Afghanistan against the Mujahideen. Following the break-up of the Soviet Union, the T-72 saw action with the Russian Army in Chechnya, Georgia and Ossetia. This, though, was in a counter-insurgency role for which it was never intended. The Russian Federation is believed to have inherited around 5,000 T-72s and a smaller number of T-80s. The T-90, a development of the T-72BM, did not go into production until after the demise of the Soviet Union so is outside the scope of this book.

While Poland and Czechoslovakia produced licensed T-72s that were built to a

An Indian-built T-72M1 Ajeya which went into production at the Heavy Vehicles Factory Avadi in 1987 – the first 175 tanks were built from Russian supplied kits and then the factory went over to local manufacture.

better standard, they lacked the resin-embedded ceramic layer inside the glacis armour and the front turret. The Polish T-72s had thinner armour and the Russian, Czech and Polish versions suffered a lack of compatibility in parts and machine tooling. India undertook local production of the T-72M1 to equip its army in the late 1980s and early 1990s.

The Yugoslavs also produced their own version, dubbed the M84, which they exported to Kuwait. Likewise, the Iraqis produced a copy known as the Asad Babil

A Polish-built 'Wolf' T-72 on the firing range. Having previously produced the T-34 and T-54/55, by April 1993 Poland had built 1,610 T-72M1s. Polish T-72s were supplied to Iran during the 1990s.

Unlike the T-62, the T-72 proved very popular with the Warsaw Pact armies – this particular one belongs to the Romanian Army. The Romanians built a series of medium tanks based on the T-54/55 and imported limited numbers of T-72s.

or Lion of Babylon – these were really kits put together from Russian spares which evaded UN sanctions. Subsequent versions included the Polish PT-91 Twardy and the Russian T-90.

T-80 MBT

The T-80, also armed with a 125mm 2A46 smoothbore gun, was accepted into service just four years after the T-72. Like its predecessor it drew on the design features of the T-64 but was greatly improved. The layout of the T-80 is generally similar to the T-64 with the driver's compartment at the front, two-man turret in the middle and engine and transmission in the rear. There are, however, many differences of detail.

The T-80's rear hull top is different to the T-64's in that it has a distinctive oblong exhaust outlet at the back of the hull. Its tracks are also wider. The T-80 reverted to the torsion-bar suspension with six forged steel-aluminium rubber-tyred road wheels and five return wheels either side, with the drive sprocket at the rear and idler to the front. There are distinctive gaps between the second and third, fourth and fifth and fifth and sixth road wheels.

The initial T-80 (1976) utilised the T-64A turret, but this was replaced by the T-80B

An early intelligence photograph of the T-80 on manoeuvres issued by the US Department of Defense. The T-80 was followed by the T-80B and the T-80U. This tank proved nowhere near as successful as the T-72.

The T-80BV's explosive reactive armour mounted around the turret and hull that appeared in 1985 posed a challenge to Western anti-armour technology.

The T-80 went into production in the late 1970s and the overall layout was similar to that of the flawed T-64, though there were many detailed improvements. Like the T-64, production of the T-80 was costly and came to a halt in 1990 except for export orders.

Export customers for the T-80U included China and Cyprus but these were delivered after the collapse of the Soviet Union.

(1979) which was also equipped with the AT-8 'Songster' missile as featured on the T-64B. This was followed by the T-80BV (1985) featuring the T-64's explosive reactive armour and the much improved T-80U (1985).

The T-64 and its successor the T-80 proved simply too costly to produce in large numbers. As the T-80 let itself down badly during the First Chechen War, the T-72 was replaced by the later T-90 (which has also been widely exported). By the late 1980s the Soviet Union had some 9,000 T-72s but just 2,500 T-80s. When the Soviet armed forces finally broke up, the Russian Federation had about 3,500 T-80s and Ukraine 345.

The T-90, a development based on the T-72BM and utilising some of the advanced features of the later-model T-80s, did not go into production until after the dissolution of the Soviet Union. It featured new-generation explosive reactive armour and the TshU1-7 Shtora-1 jammer fitted either side of the main gun.

By 1996 only 100 T-90s had been built.

A T-90 undergoing a snorkel test. Such capabilities were standard on all previous Soviet tanks.

A T-90A on parade in Moscow in 2013 – this tank has enjoyed nowhere near the level of success of the iconic T-54, T-62 and T-72.

This naval infantryman stands by his PT-76 – this amphibious light tank first appeared in the 1950s and remained in production into the late 1960s. Once open, the oval turret hatch can be locked in the vertical position.

Chapter Four

PT-76 Amphibious Light Tank

During the Cold War the Soviets took their amphibious assault forces very seriously. Although over half a million Soviet sailors had fought ashore during the Second World War, the Soviet fleet did not revive its naval infantry forces until the mid-1960s. These numbered about 20,000, though in the event of war they would probably have been mobilised to three times this number. Moscow considered them an elite force with the men receiving airborne, artic and mountain warfare training.

The Soviets, supported by their Warsaw Pact allies, ensured they had significant amphibious capabilities in the Baltic and Black Seas. Likewise, Soviet naval infantry

The massive 'Aist' class assault hovercraft could carry four PT-76s and fifty troops, two T-54/55s with 200 troops or three APCs and 100 troops. During the Cold War such amphibious assault forces operated in the Baltic Seas and the Pacific Ocean.

operated with the Soviet Pacific Fleet and also with the navy's river flotillas. A typical naval infantry brigade was equipped with medium and light tanks as well as wheeled APCs and were transported by the 'Alligator' class landing ship tank and 'Aist' class assault hovercraft.

At the forefront, supporting their naval infantry, was a distinctive amphibious tank. The Soviets knew that this would have a role in forcing Europe's major rivers should war break out in central Europe. The PT-76 *Plavayushchiy Tank* or amphibious tank was first accepted into service in 1950 after being developed by the IV Gavalov OKB Design Bureau as the K-90. Around 7,000 of these light tanks had been built by 1967 when production finally came to an end. While it never saw combat with the Soviets, it was involved in some fierce fighting during numerous Cold War proxy conflicts.

Interestingly, the PT-76 shared its heritage with some illustrious predecessors. The tank was armed with the 76.2mm D-56T gun which was a development of the weapon used by the T-34/76 and the KV-1 tanks during the Second World War. It had a maximum rate of fire of between six and eight rounds a minute with a

This grainy propaganda photograph shows a PT-76 coming ashore during a Soviet naval exercise. The tank is propelled through the water by two rear-mounted water jets.

This tank is on display in Kiev's Museum of the Great Patriotic War. Its 76.2mm D-56T gun is fitted with a double-baffle muzzle brake and fume extractor.

maximum range in the indirect fire role of around 13,000m. In addition a 7.62mm SGMT machine gun was mounted coaxially to the right of the main armament. Many were also fitted with a 12.7mm DShKM anti-aircraft machine gun.

The tank's hull was of welded steel and was divided into three compartments, with the driver to the front, fighting compartment in the middle and the engine in the rear. Unusually, the driver was seated centrally with a single-piece hatch that swung to the right. The turret was of all-welded steel with the commander, also acting as the gunner, seated on the left with the loader on the right. It had a single oval-shaped hatch that hinged forward and could be locked vertically. To the left of the hatch was a circular cupola which housed three integral periscopes and could be manually traversed 360 degrees by the commander. The commander also had an optical TSh-66 sight to the left of the main gun while the loader had a periscope mounted in the turret roof forward of the hatch. The driver was also served by three periscopes which were mounted forward of his hatch cover.

The first production model was armed with the D-56T gun fitted with a multi-slotted muzzle brake. The subsequent and more common model was fitted with a double-baffle muzzle brake and a bore evacuator towards the muzzle. The PT-76B was fitted with a fully stabilised D-56TS and also had the benefit of a nuclear, biological and chemical (NBC) protection system. The Model V-6 engine used in the PT-76 was one bank of that fitted to the T-54. The manual gearbox had five forward and one reverse gears and steering was of the clutch and brake variety. The torsion bar suspension comprised six road wheels with the drive sprocket at the rear and the idler at the front. The first and sixth road wheel stations had hydraulic shock absorbers and the steel tracks consisted of ninety-six links.

This North Vietnamese Army PT-76 is one of ten knocked out during the Battle of Ben Het by US M48 Pattons on 3 March 1969. The PT-76 was widely exported but only in small numbers, though the NVA were supplied with about 250.

A BRDM-1 amphibious scout car and PT-76 light tank on exercise with Polish naval forces. A key unit deploying this tank was the Polish 7th Coastal Defence Brigade, which at one stage had a complete battalion of sixty PT-76s.

The tank was propelled through the water by two water jets mounted to the rear. To enter the water all the crew had to do was to erect the trim vane at the front of the hull and activate two electric bilge pumps; the later were backed up by an emergency manual bilge pump. Steering the tank whilst in the water was simply done by opening and closing two hatches over the water jets.

A typical 3,000-strong naval infantry brigade was equipped with forty-four PT-76s, a similar number of T-72 tanks (which started to replace the navy's T-55s in the mid-1980s) and 145 BTR-60/70 APCs. The 'Aist' class hovercraft was capable of lifting four PT-76s, one T-72 or 220 troops. A key unit was the Red Banner Northern Fleet's 63rd Guards Naval Infantry Brigade, based at Pechenga on the Kola Peninsula. Its main role was to spearhead any invasion of Norway or Iceland.

Soviet Army tank and motor rifle regiments' reconnaissance companies were equipped with five PT-76s and three APCs, while each division had a separate reconnaissance company deploying a further five PT-76s. However, this light tank's limited utility meant that it was gradually replaced by T-54/55s, T-62s and T-72s and even BMP-1/2s IFVs. Because it was amphibious early models had no NBC protection, night fighting equipment and very thin armour.

Nonetheless the PT-76 saw extensive combat around the world including Africa, the Middle East, in the Indo-Pakistan conflict and during the Vietnam War. In fact, its most iconic moment occurred when the North Vietnamese used it to overrun the

US Special Forces base at Lang Vei in 1968. The PT-76 ended up in the service of at least twenty countries, though Vietnam and Iraq were by far the largest recipients, receiving 250 and 100 respectively. The Chinese also produced a version known as the Type 63 but this has a very different cast turret. When the Soviet Union collapsed, the Soviet armed forces still had several hundred PT-76s and a number were serving with the Polish 7th Coastal Defence Brigade.

It is believed that the PT-76 was manufactured at the Kirov facility in Leningrad and the Volgograd Tractor Factory. Notably many of this light tank's components were also used with the BTR-50 APC, the SA-6 'Gainful' SAM system and the ZSU-23-4 self-propelled anti-aircraft gun.

Chapter Five

BTR Wheeled Armoured Personnel Carriers

BTR-152

The BTR-152 6x6 was developed after the Second World War as the Soviet Union's very first purpose-built APC. It was manufactured in large numbers from 1950 and saw service with African and Asian armies. The all-welded steel hull showed close similarities with American and German wartime designs. Notably, significant numbers of the M3A1 4x4 scout car and M2 and M5 series of American half-tracks were supplied to the Red Army under Lend-Lease arrangements. Likewise the Soviets captured large numbers of the Hanomag-built range of German half-tracks.

As with these earlier vehicles, the BTR had a front-mounted engine and an open top crew compartment for the driver and troop compartment for up to seventeen soldiers. The driver and commander had separate glass windscreens that could be protected by steel hatches with vision blocks. The infantry entered and exited the vehicle either via the open roof or through a single door in the rear plate of the hull. For defensive purposes the vehicle had six firing ports, three either side and two in the rear plate either side of the door.

The BTR-152 was the Soviet Union's very first purpose-built APC, utilising a truck chassis. It was widely exported throughout Europe, Africa, the Far East and the Middle East.

Known as the Model D or M-1961 the BTR-152K had full overhead armoured protection with two roof hatches over the troop compartment.

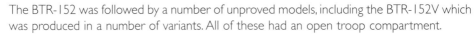

The BTR-152 was followed by a number of unproved models, including the BTR-152V which was produced in a number of variants. All of these had an open troop compartment.

Initially the ZIS-151 2½-ton 6x6 chassis was used as the basis for the BTR-152, though later models utilised the ZIS-157. The six-cylinder, inline model ZIS-123 was a water-cooled petrol engine generating 110hp at 2,900rpm. The BTR-152's transmission layout was that of a conventional 6x6 commercial truck with the drive shafts leading to differentials on 'solid' axles. The gearbox had five forward speeds and there was a two-speed transfer box. The tyres had a pressure system regulated by the driver to suit the ground conditions. Some BTR-152s also featured a front-mounted winch.

Some versions were fully enclosed, such as the BTR-152U command variant, which has much higher sides to allow staff officers to stand up inside. The normal armament comprised the standard 7.62mm machine gun or the heavier 12.7mm or 14.5mm mounted on the hull top. The BTR-152A-ZPU was an anti-aircraft variant armed with twin 14.5mm KPV machine guns in a rotating turret. Against aerial targets, these were only effective to 1,400m. They also carried AP rounds for use against light armoured vehicles, which could penetrate 32mm of armour at 500m, though the guns had a range of 2,000m against ground targets. Other anti-aircraft variants included the BTR-152D and the BTR-152E.

Some of those supplied to the Egyptian Army were armed with the Czech quad 12.7mm M53 anti-aircraft system. This comprised four Soviet 12.7mm DShKM machine guns on a Czech-designed two-wheel mount. A number of these ended up in service with the Afghan Army. Likewise, in 1982 the Israeli Army encountered BTR-152s being operated by the Syrian-backed Palestinian Liberation Army that

This photograph shows the BTR-152 with the windscreen and door shutters open. The roof over the commander and driver's position is armoured.

were fitted with a twin 23mm automatic anti-aircraft gun in the rear of the troop compartment.

BTR-40

The BTR-152's smaller cousin was the BTR-40, introduced in 1951. This was essentially a redesigned version of the American-supplied M3A scout car. It was based on the GAZ-63 truck chassis, but with a shorter wheelbase and was a conventional four-wheel drive armoured truck with a frontal engine layout. In the event of chemical

The BTR-40 was essentially the BTR-152's smaller cousin with a 4x4 chassis. This one belonged to the Egyptian Army and was captured by Royalist guerrillas during the North Yemen Civil War in the 1960s.

warfare one variant of this vehicle was designed for a chemical decontamination role, which included placing flag markers to warn of contaminated areas. A more conventional version was the BTR-40A/ZPU; this had an anti-aircraft role mounting twin 14.5mm KPV heavy machine guns. These were mounted in a manually-operated open turret with a 360-degree traverse and an effective rate of fire of 150 rounds per minute.

BTR-60

The requirement to replace the non-amphibious BTR-152 was issued in the late 1950s, and the heavy eight-wheeled amphibious BTR-60P entered service with the Soviet Army in 1961. Since then it has been supplied to armies throughout the world and was built in Romania as the TAB-72. The BTR-60P was powered by two GAZ-49B six-cylinder, water-cooled, in-line petrol engines, developing a total of 180hp. These were mounted in the rear of the welded steel hull and drove all eight wheels, the front four of which were steerable. The BTR-60 series was fully amphibious, propelled through the water by a hydrojet system with a single controllable outlet at

The BTR-60 entered service in 1961 and appeared in three main versions, the BTR-60P (seen here), the BTR-60PA and the BTR-60PB.

The drawback with the initial BTR-60 was its open troop compartment meant the occupants were vulnerable to indirect fire

The BTR-60A, which had overhead armour protection, was also known as the BTR-60PK.

The driver's station in the BTR-60PB. Like those of all Soviet armoured fighting vehicles it is very cramped. The Romanians built the BTR-60PB as the TAB-71.

the rear. This gave a calm-water speed of 10km/h compared to 80km/h on land. During deployment in water a bilge pump was available, together with a trim vane that was normally carried flat on the nose plate.

The troop compartment (initially for fourteen men but reduced in later models) occupied the centre of the vehicle with the driver on the left and the commander on the right at the front. The troop compartment had no overhead protection but this was remedied with the BTR-60PA or BTR-60PK, which was fully-enclosed with roof hatches, installed to supplement access through two small hatches on each side.

The final model, the BTR-60PB, was fitted with a small turret on the hull roof near the front, mounting a 14.5mm machine gun and a 7.62mm machine gun. It is identical to that fitted to the Soviet BRDM-2 reconnaissance vehicle and the Czech OT-64 APC. While the BTR-60PB was built under licence in Romania as the TAB-71, the lack of easy access resulted in the Czech and Polish governments developing the SKOT (OT-64) series for their armies. Production of the BTR-60 series ended in 1976, resulting in around 25,000 vehicles.

The BTR-60PB was essentially the 60PA with a machine-gun turret and other modifications. This version proved highly popular and was supplied to numerous armies, but the Warsaw Pact armies sought to improve on the vehicle by developing their own designs known as the SKOT and TAB.

On the BTR-60 the large forward-opening side hull hatch necessitated spreading out the firing ports. On the subsequent BTR-70 this hatch was placed lower down between the second and third axles with the firing ports clustered above it.

This BTR-60PB was caught in an ambush in Afghanistan in the 1980s. This vehicle only really provided the occupants with protection for from small-arms fire and was easily disabled or knocked out by the Mujahideen.

The rail-type antenna mounted on this BTR shows that it is a command vehicle. The exposed front wheel indicates it is a BTR-60 PU-12 normally associated with air-defence units.

BTR-70

The follow-on BTR-70 first appeared during the November 1980 military parade in Moscow. The hull was of all-welded steel armour with improved protection over its front arc compared to the BTR-60. In addition the nose was wider and the front gave added protection to the front wheels. While the BTR-70 was fitted with the same turret as its predecessor, some were fitted with the BTR-80 turret. Initial models of the BTR-70 were fitted with the same wheels and tyres as the BTR-60.

The two GAZ-49B engines were replaced by two ZMZ-4905 petrol engines, which developed 120hp each compared to just 90hp each in the BTR-60. Both engines had their own transmission with the right engine supplying power to the first and third axles, while the left powered the second and fourth axles. This meant if one engine was out of action the vehicle could still move, albeit at a slower speed. The exhausts were less boxy than on the BTR-60. Whereas the BTR-60 could carry up to sixteen men, the BTR-70's capacity was two crew and nine passengers. Again Romania produced its own version, dubbed the TAB-77.

A BTR-70 on operations in Afghanistan. It first appeared publicly in 1980 but production actually ran from 1972 to 1982. Large numbers of BTR-60/70/80s were supplied to the Afghan Army.

The lower hull hatch and firing port configuration is clearly visible on this preserved BTR-70.

Although the BTR-70 was an improvement over the earlier BTR-60, it still had its problems, not least the inadequate means of entry and exit for the troops and the two petrol engines which were inefficient and could catch fire. The Soviet Army first took delivery of the improved BTR-80 in 1984.

BTR-80

A key difference in appearance of the BTR-80 was that a new hatch was installed between the second and third axles; the upper part of this opened to the front while the lower part folds down to form steps, permitting troops to dismount much more quickly and with less exposure to enemy fire than in the earlier BTRs.

In addition, while the BTR-70 had three firing ports in either side of the troop compartment, the BTR-80 had its three firing ports angled to fire obliquely forward, thereby giving covering fire into the dead ground towards the front of the vehicle. There·was also a single firing port to the right of the commander's bow position that was also on the BTR-70, plus an additional firing port in each of the two roof hatches.

The two forward firing ports were for the 7.62mm PK general-purpose machine gun, while the three firing ports either side could accommodate the AKMS/AK-74 Kalashnikov assault rifle. Small arms carried by the crew consisted of two 7.62mm machine guns and eight 7.62mm AKMs or 5.56mm AK-74s and nine Type F1 hand-grenades. For air defence they also normally carried two man-portable surface-to-air missiles such as the SA-14, SA-16 or SA-18. Less visible was the replacement of the two petrol engines with a single V-8 diesel engine developing 260hp, which provided a significant increase in power-to-weight ratio. This meant a slightly improved road speed and better fuel efficiency along with reducing the risk of fire.

While the one-man manually-operated turret was similar to that fitted to the BTR-70 and the BTR-60PB, the 14.5mm KPV heavy machine gun had twice the elevation of the earlier models. This meant it could act in an air-defence role against low-flying aircraft and helicopters. Another visual difference with the earlier versions was that mounted on the rear of the turret of the BTR-80 was a bank of six electrically-operated smoke grenade dischargers (81mm Model 902V). This was operated from within the turret and each grenade could generate a smoke screen of up to 30m wide and 10m high.

This BTR-80 was photographed in Afghanistan – it is readily identifiable by the horizontal exhaust on the rear deck (on the BTR-60 and 70 the exhausts sloped downwards). Within the Warsaw Pact, only Hungary took receipt of the BTR-70/80.

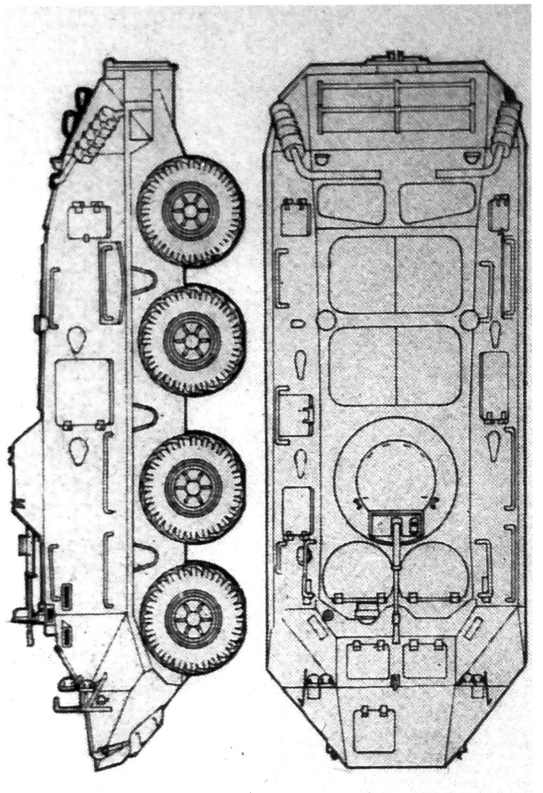

This Soviet drawing of the BTR-60PB highlights some of its drawbacks. In the later models the front of the hull was extended to offer greater protection for the front wheels, while the side hull hatch was placed in the lower hull. The upper deck hatches also exposed troops leaving the vehicle and this was only remedied with the advent of the BMP which has a door in the rear.

The smoke grenade launchers are just visible on the rear of this BTR-80's turret.

Essentially the BTR-80 and its predecessors were little more than armoured 'battle buses' that provided protection from heavy machine guns only out to 100m. In the case of the BTR-80, on the frontal arc the armour gave protection against 12.7mm AP rounds at a range of 100m, the upper hull only provided protection against 7.62mm AP at 100m while the lower hull can withstand 7.62mm AP at 750m.

By the mid-1990s the Russian Army was estimated to still have almost 10,000 BTR-60/70/80s. The BTR80A (which has a turret armed with a 30mm 2A72 cannon) did not appear until after the Soviet Union collapsed. The BTR-152 and BTR-60/70/80 range of APCs were widely exported around the world and were involved in innumerable bush wars.

Chapter Six

BMP Infantry Fighting Vehicle & MT-LB Armoured Personnel Carrier

During the Second World War the Red Army lagged way behind its Western counterparts in producing adequate APCs for its troops. While Western armies made use of armoured half-tracks, turretless tanks and de-gunned self-propelled guns to transport infantry across the battlefield, the Soviets had largely relied on unarmoured lorries and infantry riding on the outside of tanks.

BTR-50 APC

In 1973 the Egyptian Army stormed across the Suez Canal to drive the Israeli armed forces from the Sinai desert. Amongst those armoured vehicles photographed rumbling over the pontoon bridges was the Soviet-supplied BTR-50 amphibious tracked APC. This had originally been developed as a hurried response to the first generation of Western APCs, which had begun to appear in Central Europe in the late 1950s. First seen in 1957, the open-topped BTR-50P was based on the PT-76 amphibious tank chassis.

The BTR-50 was designed to transport an infantry section or artillery up to 85mm calibre that could be fired from the vehicle without offloading. The vehicle required two crew and could carry up to ten soldiers. Predictably, the open variant left the passengers exposed and this was quickly followed by a version known as the BTR-50PK that had an armoured roof. Other variants included the BTR-50PA armed with a 14.5mm KVPT or ZPU-1 machine gun, the BTR-50PU command vehicle and some special-purpose versions designed to carry electronic countermeasures equipment.

The BTR-50 was supplied to the Soviet and Warsaw Pact armies and was exported outside Europe including the Middle East. However, the BTR-50 suffered a major shortcoming in that it had no rear doors, which meant that troops had to debus via roof hatches and over the sides with predictable results when under fire.

The initial model BTR-50P amphibious tracked APC had an exposed open-topped troop compartment. This vehicle was based closely on the amphibious PT-76 chassis.

The subsequent BTR-50PK had an armoured roof over the troop compartment created by two large rectangular doors that open to the outside of the vehicle.

The major disadvantage with the hastily-conceived BTR-50 was that the troops it was transporting had to debus through the roof hatches, which made them very vulnerable when under fire.

This BTR-50PK belonged to the East German Army. Although replaced in frontline service by the BMP, thousands of BTR-50s saw service with all Warsaw Pact armies.

The Czechs produced a modified version of the BTR-50PK known as the OT-62 (seen here), which had a revised front compartment plus a longer range and higher road speed. In appearance it is almost identical to the Soviet BTR-50PU Model 2 Command Vehicle.

As a result it was replaced by the BMP IFV. Nevertheless it still saw combat with, amongst others, the Egyptian and Iraqi armies.

The Czechs and Poles produced a modified version of the BTR-50PK dubbed the OT-62. This had an improved road speed, greater range and a redesigned front compartment. The standard Polish version was designated the TOPAS 2AP, which along with the PT-76 amphibious tank were deployed by the 'Polnocny' class landing ships tank in support of Warsaw Pact naval assault forces.

BMP-1 IFV
One of the most innovative and revolutionary armoured vehicles to emerge from the Cold War was the tracked *Boevaya Mashina Pekhota* (BMP) IFV; this predated the American Bradley, the British Warrior, the French AMX-10 and the German Marder tracked IFVs. The BMP represented the first true mechanised IFV – essentially a hybrid APC and tank. It was designed as a breakthrough vehicle intended to help the Warsaw Pact cut its way through Central Europe. Its key role was to ensure the swift and mass exploitation of a breakthrough of a lightly-defended point in support of the

The BMP-1, which appeared in the mid-1960s, helped revolutionise mechanised warfare. At this stage the Americans were only just beginning to experiment with armoured cavalry by adding machine guns to the M113 APC. In contrast the BMP-1 had a turret-mounted 73mm 2A28 gun. This particular vehicle has the launch rail bracket but no AT-3 'Sagger' anti-tank missile. The 'casualty' is being removed via the commander's hatch.

infantry. In theory the BMP was to charge forward, guns blazing, before disgorging its infantry to seize and hold enemy ground. Its 73mm gun and anti-tank missiles were designed to ensure that it could engage enemy tanks should the need arise.

The BMP-1 first appeared at the November 1967 Moscow military parade, but this is believed to have been a pre-production model. It was followed by the initial production run known as the BMP Model 1966, though the main production variant is thought to be the Model 1970. The BMP first saw combat with the Egyptian and Syrian armies during the 1973 conflict with Israel. Since then it has seen action with the Soviet Army in Afghanistan, with government forces in Angola, with the Iraqi Army and with the Libyan and Syrian armies.

The vehicle consisted of an all-welded steel hull with a distinctive ribbed and sloping glacis plate. The driver was positioned at the front on the left and had a single hatch cover that opened to the right. He was served by three periscopes; the central TNPO-170 could be replaced by the TNPO-350B which was vertically extendable and allowed the driver to see over the trim vane when it was erected for amphibious operations. The troop compartment in the rear could transport eight soldiers seated back-to-back with four down each side. Access was via two rear doors (that each housed integral fuel tanks and a vision device, and the left door had a firing port), or via four roof hatches.

The one-man turret was the same as that on the BMD-1 airborne combat vehicle. The gunner had a single hatch that opened forwards, in front of which to the left side was located a dual-mode IPN22M1 monocular periscope sight. Four

An early intelligence photograph of a BMP-1. The all-welded steel hull provided the crew with protection from small-arms fire and shell splinters.

BMP-1s night-firing their 73mm main armament.

This column of BMP-1s and MT-LBs were photographed somewhere in Central Asia, quite possibly Afghanistan.

additional observation periscopes served the turret gunner and a white light or infra-red searchlight was mounted on the right side of the turret. Because it had a turret the public and media often mistook the BMP for a tank. This housed the main armament; the 73mm Model 2A28 smoothbore, low-pressure, short-recoil gun which weighed 115kg. This was served by a forty-round magazine located to the right rear of the gunner.

The gun fired a fixed, fin-stabilised PG-9 HEAT projectile which employed a small PG-15P stub casing to boost the projectile out of the barrel at an initial velocity of 440m/s at which point the PG-9V rocket motor ignited to supply the main source of propulsion, accelerating the shell to 700m/s. This was the same round as that used in the SPG-9 infantry weapon and had a maximum effective range of 1,300m. The projectile could overcome up to 300mm of armour.

Mounted over the main gun was an AT-3 'Sagger' wire-guided anti-tank missile launcher. A single missile was carried ready to fire with two reloads in the turret, which were loaded via a rail through a hatch in the forward part of the turret roof.

Soviet tank crew on exercise. The mechanised infantry they are conferring with are acting in support in the BMP-1 parked directly behind them.

A Soviet Army BMP-1 on exercise – the 'Sagger' anti-tank missile is just visible mounted above the main armament. The Soviet Union amassed 70,000 armoured fighting vehicles at the height of the Cold War, which included some 24,000 BMP-1/2s.

Mounted coaxially to the main gun was a 7.62mm PKT machine gun, fed by a continuous belt of 2,000 rounds held in a honeycombed ammunition box mounted below the weapon. The turret traverse and gun elevation were electric, with backup mechanical controls in case of power failures.

The BMP's torsion bar suspension was made up of six rubber-tyred road wheels either side, with the drive sprocket at the front, the idler at the back and three track-return rollers. The first and last road wheel stations had a hydraulic shock-absorber and the top of the track was protected by a light sheet-steel cover. The track links were the double-pin type with water scoops between the housings. The BMP-1 was fully amphibious and was propelled through the water by its tracks. Variants of the BMP-1 were built by China, Czechoslovakia and Romania.

While the Soviet Army had high expectations of the BMP, they came in for an unpleasant surprise in 1973. During the Arab-Israeli Yom Kippur War the Egyptians used the BMP exactly as the Soviet manual dictated. Soviet theory was all well and

Polish BMP-1s on exercise. By the mid-1980s Poland had about 800 BMP-1s in its order of battle which were supported by about 2,500 wheeled SKOT and TOPAS armoured personnel carriers.

A concealed BMP-1 on exercise with the Soviet Army during the 1980s.

good but in the open tank ground of the Sinai the BMP proved ultimately to be too vulnerable to Israeli tanks, anti-tank weapons and jet fighters. The 73mm gun proved to be largely ineffective and the missile was difficult to control. To compound matters, Egyptian crew training was probably not as good as it could have been.

Soviet doctrine had to be rethought, resulting in the infantry dismounting about 300m from their objective, which was to be taken on foot under covering fire from the BMP gunner, supporting tanks and artillery. Despite this doctrinal rethink, in the intervening years the BMP has seen combat in numerous wars around the world. Despite its limitations against enemy tanks, it provided infantry with a welcome force-multiplier and set a trend that NATO followed.

BMP-2 IFV

Lessons learned from the BMP-1 inevitably led to a BMP-2. This first appeared in November 1982 in the Red Square parade, though it is believed to have already been in service for a number of years before that. While visually the BMP-2 is almost identical to its predecessor, a clear difference is the long thin barrel of the main armament that consists of the 30mm Model 2A42 cannon. This is housed in a two-man all-welded steel turret with the commander seated on the right and the gunner on the left. The gunner has a single rectangular hatch, which opens to the front with an integral rear-facing periscope and three fixed periscopes, with two to the front and one to the left side. A total of 500 rounds are carried for the main gun.

In addition the BMP-2 has an AT-5 'Spandrel' anti-tank missile launch tube mounted on the turret roof between the gunner and commander's hatches. As well as the infantrymen's small arms, the BMP-2 also normally carried an anti-tank grenade launcher and two surface-to-air missiles. The infantry compartment at the rear only has two roof hatches compared with the four fitted on the BMP-1, though access is normally via the two rear doors. It only carries six infantrymen compared to eight in the BMP-1.

Like its predecessor, the BMP-2 is fully amphibious. Just before entering the water a trim vane stowed on top of the glacis plate is erected, the bilge pumps are switched on and the driver's centre periscope is replaced by the TNPO-350B. The upper part of the tracks has a sheet metal covering that is deeper than that on the BMP-1 as it is filled with a buoyancy aid.

The follow-on BMP-2 is easily identifiable by the long, thin barrel of its 30mm 2A42 cannon. The larger two-man turret meant that the vehicle carries fewer infantry than the BMP-1 and the infantry compartment only has two roof hatches rather than four.

An Iraqi BMP-1 captured during Operation Desert Storm in 1991. By the mid-1980s Iraq had about 500 BMPs.

This burnt-out Iraqi BMP-1 was destroyed during Desert Storm.

From the late 1980s onwards a number of enhancements were carried out to production BMP-2, most of which were retrofitted to earlier BMP-1 and BMP-2. The latter was supplied to the Iraqi Army and was manufactured in India as the Sarath and in Czechoslovakia as the OT-90. The BMP-3, which features redesigned road wheels and a higher hull profile, appeared just as the Soviet Union was collapsing. This is an upgunned BMP that has a turret-mounted 2K23 weapon system that comprises a 100mm 2A70 gun, a coaxial 30mm 2A72 cannon and a coaxial 7.62mm machine gun.

MT-LB MTV

The versatile MT-LB, by contrast, started life in the late 1960s as an armoured multipurpose tracked vehicle. In all, there have been more than eighty variants but its main roles were as an artillery prime mover, cargo carrier and an armoured personnel carrier. Although slightly smaller than the BMP it carried more men, three crew and ten infantry. Armour protection was, however, rather less and the main armament was limited to one 7.62mm machine gun.

The MT-LB was designed as a multi-purpose fully amphibious auxiliary armoured tracked vehicle that, despite its shortcomings, proved highly successful and was very popular with the Warsaw Pact armies.

The MT-LB's hull was all-welded steel with the crew compartment at the front and the engine immediately behind and the troop compartment at the rear. As with most Soviet armoured vehicles, the driver sat on the left and had a single-piece hatch in front of which were three periscopes. The commander was located to the right of the driver and was served by a single-piece hatch and two periscopes. When in combat the commander also operated the turret.

The MT-LB was initially dubbed the M1970 by NATO until its correct Soviet designation became known.

Former East German Army MT-LBs were used by the US Marines for training purposes.

The machine-gun turret was mounted above the commander's position and was armed with a 7.62mm PKT machine gun. Like the turrets on the BRDM-2 and BTR-60PB it did not have a hatch cover. The machine gunner and the driver both had a windscreen in front of them, which when in battle was protected by an armoured flap hinged at the top. An aisle gave access from the front crew compartment to the personnel compartment at the back. The latter had inward-facing folding canvas seats for ten infantrymen. Two hatches over the top of the troop compartment opened forwards. Troops entered and left the MT-LB by two doors in the rear of the hull, both of which had a firing port. On each side of the troop compartment there was an additional firing port and vision block.

The standard torsion-bar suspension consisted of six road wheels with the idler at the rear and the drive sprocket at the front. There were no track return rollers as the tracks rested on top of the road wheels. The vehicle was fully amphibious, propelled through the water by its tracks. Like most armoured vehicles, just before entering the water a trim vane was erected at the front of the MT-LB and the bilge pumps were switched on.

There have been over eighty variants of the MT-LB, including the stretched MT-LBU that has a higher hull, longer chassis and features seven road wheels rather than six. By the early 1990s the Russian Federation and Ukraine still had some 4,000 MT-LBs.

An Iraqi MT-LBU captured in 1991 during Desert Storm. Iraq obtained about 500 MT-LBs, many of them built in Bulgaria.

A Bulgarian MT-LB variant known as the BMP-23 first appeared in the early 1980s. This was armed with a 23mm cannon and was followed by the BMP-30 armed with a 30mm cannon – but only small numbers were ever produced.

The MT-LB was not produced in such large numbers as the BMP but entered service with the Soviet and other Warsaw Pact armies. It was also manufactured in Bulgaria and Poland for many years. Notably Bulgaria built it under licence for both domestic and export markets that included Iraq. Bulgarian variants included the MT-LB mortar carrier which could carry either an 82mm or 120mm mortar in the rear of the hull.

Soviet airborne troops on exercise supported by BMD-1s, none of which are carrying the 'Sagger' missile. Such photographs regularly appeared in *Red Star*, the Soviet military newspaper, and were intended to emphasise the power of Moscow's seven airborne divisions and eight air assault brigades.

Chapter Seven

ASU Airborne Assault Vehicle & BMD Airborne Combat Vehicle

On 21 August 1968, a vengeful Moscow moved to crush the Prague Spring. Spearheading the 103rd Guards Air Assault Division, transport aircraft disgorged a number of ASU-85 airborne assault vehicles and APCs onto the runway at Ruzyme airport on the north-east outskirts of the Czech capital. These then seized the Presidential palace on Hradcany Hill in Prague; other key locations

The ASU-85 was air portable and could be dropped by parachute. It was used to spearhead the invasion of Czechoslovakia in 1968.

were also taken. Two regular Soviet motor rifle divisions then reinforced the airborne forces. The ASU-85s were soon on the streets of Prague, cowing the rebellious population. The Czechs were so surprised by their sudden appearance that there was hardly any opposition.

Moscow repeated this success in 1979 when elements of the 105th, 104th and 103rd Air Assault Divisions captured the Afghan capital Kabul. The airborne forces allotted to this operation were larger as it was anticipated resistance would be much more widespread. Unlike the 1968 operation, this time the airborne forces were fully mechanised with BMD airborne combat vehicles that were used to occupy Bagram air base outside Kabul. The airborne task force used to assault the Darulaman Palace was reportedly equipped with BMDs and ASU-85s which took on Afghan tanks.

ASU-85 Airborne Assault Vehicle

Following the Second World War the Soviets developed considerable airborne forces, known as the VDV (*Vozdushno Desantnaya Voyska* – Air Assault Force), which were supported by a series of AFVs that were air-portable and could be dropped by parachute. Key amongst these were the ASU and BMD. The successor to the much smaller and less powerful ASU-57, the ASU-85 (*Aviadesantnaya Samokhodnaya Ustanovka* – airborne self-propelled mount) appeared in public for the first time in 1962 and was the Soviets main airborne assault vehicle. Only 2.1m in height, the ASU-85 could be transported by air or parachute-dropped. Each Soviet airborne division had an assault gun battalion equipped with thirty-one ASU-85s and the Polish 6th Pomeranian Airborne Division also deployed it.

Based on the PT-76 tank, the ASU-85 had the same engine, transmission and running gear and was roughly the same weight. It was not amphibious, having been adapted to the assault gun and tank-destroyer role. The 85mm 2A15 gun, which fired HE as well as AP rounds, was located just left of centre of the sloping glacis plate and had a traverse of 12 degrees and elevation of 15 degrees. The driver sat to the right of the main gun, the other three crew members, commander, gunner and loader behind. The vehicle carried forty-five rounds for the main gun and 2,000 rounds for the coaxial 7.62mm machine gun. The TShK-2-79 daytime or the TPN1-79-11 night sights directed both the 85mm and the machine-gun.

The only upgrade, dubbed the ASU-85 M1974 by NATO, appeared in the early 1970s. This simply consisted of the installation of a DShk-M 12.7mm heavy machine gun with 600 rounds, to give the vehicle some measure of anti-aircraft defence. This meant that the ammunition load for the main armament was reduced to thirty-nine rounds.

The ASU-85 airborne self-propelled gun was based on the PT-76 amphibious tank and was armed with a 85mm gun mounted in the hull. This particular one belonged to the Polish 6th Air Assault Division.

BMD-1 Airborne Combat Vehicle

The BMD-1 (*Boyevaya Mashina Desantnaya* – airborne combat vehicle) first entered service with Soviet airborne units in 1969 but was not seen publicly for another four years. Its main claim to military fame is that it spearheaded Moscow's invasion of Afghanistan in December 1979, helping to secure Kabul. Since then it has been produced in three different variants, though the basic vehicle remained the same.

This small IFV had a crew of three and could carry four other passengers. Its main armament was the same as that on the BMP-1 IFV, the 73mm Model 2A28, loaded from an automatic forty-round magazine to the right rear of the gunner. Traverse and gun elevation was electric with the usual manual controls for emergencies. The gunner was served by a dual-mode IPN22M1 monocular periscope sight mounted on the left side of the turret. Day mode magnification gave x6 and a 15-degree field of view, while night mode offered x6.7 and a 6-degree field of view.

Above the 73mm gun was a launcher rail for an AT-3 'Sagger' anti-tank missile. Two missiles were carried inside the turret, which were loaded via a rail through a hatch in the forward part of the turret. Controls for the 'Sagger' were stored under the gunner's seat. When needed these were locked in position between the gunner's legs, who controls the missile using the joystick in the usual manner. Mounted coaxially to the right of the main armament was a 7.62mm PKT machine gun, fed

The BMD-1 airborne combat vehicle was unique during the Cold War, as NATO had nothing comparable in the parachute or air-landing role.

The BMD-1 entered service with Soviet airborne units in the late 1960s and about 3,000 were built. This vehicle is just 5.4m long, 2.6m wide and 1.6m high.

from a continuous belt of 2,000 rounds loaded into an ammunition box below the weapon. To catch the spent casings a cartridge and link collector was mounted in the turret basket.

The hull of the BMD-1 was of welded aluminium. The driver was located at the front of the vehicle, seated in the centre just forward of the turret and had a single hatch that opened to the right. Three periscopes were mounted forward of the hatch. The commander sat to the left of the driver and beside the commander's seat were the radio and gyrocompass. The bow machine gunner sat to the driver's right and aimed the bow-mounted 7.62mm PKT machine guns using a TNPP-220 periscope sight. The two machine guns were mounted at either side of the front of the vehicle. Two semi-circular hatches were positioned either side of the forward edge of the turret.

This gives a good view of the BMD-1 turret: note the 'Sagger' missile mounted on a launch rail on top of the 73mm gun.

The BMD-1's main claim to fame is that it spearheaded the Soviet invasion of Afghanistan in 1979.

This Iraqi BMD-1 was destroyed in 2003 during Operation Iraqi Freedom. Iraq was only supplied a small number of these vehicles.

The BMD-1's turret had a single-piece forward-opening hatch to the left. The gunner had four periscopes; one mounted either side and two forward of the hatch. The rear personnel compartment had a concertina-style hatch which opened towards the front.

The suspension comprised five small road wheels with the drive sprocket at the rear and the idler at the front, plus four track return rollers. The suspension combined a hydraulic system for changing the ground clearance and maintaining track tension with pneumatic springs, enabling ground clearance to be changed from 100mm to 450mm. The BMD-1 was fully amphibious, propelled through the water by two water jets at the rear of the hull. Limited numbers of BMD-1s were supplied to Angola, India and Iraq.

Soviet airborne or air-landing divisions were issued with 330 BMDs per division; three command versions with the divisional headquarters and three regiments with 109 BMDs each (ten command vehicles, nine BMDs without turrets and ninety basic BMD-1s). Initially the BMD-2 was issued at a rate of nine to each of the three airborne regiments.

BMD M1979/BTR-D

This is immediately distinguishable from the BMD-1 by its longer chassis which has six rather than five road wheels and five rather than four return rollers, the lack of turret and different hull top. This vehicle was first seen during the Soviet invasion of Afghanistan and was dubbed the BMD M1979 by NATO. Development of the BTR-D commenced in 1974 drawing on the automotive parts of the BMD-1. Like the latter it had a hull of all-welded construction. The glacis protection was increased using a dual-slanted angle in the upper plates of the armour at the front.

The BTR-D was designed for a variety of roles including personnel transport, towing support weapons and maintenance support. It could carry ten infantry as well as the three crew, though the bow machine gunners also normally deployed with the infantry. Some early models were fitted with a small one-man turret armed with a 7.62mm PKT machine gun. A number were also armed with a 30mm AG-17 automatic grenade launcher. A command post variant called the BMD-KShM with a 'clothes rail' radio antenna was also seen deployed to Afghanistan.

BMD-2 ACV

The BMD-2 that went into production in the late 1980s was essentially a BMD-1 chassis with a new turret equipped with a different main gun. Initially it was assumed that the BMD-2s were simply rebuilds of the earlier model, but they were in fact new-build vehicles. While the chassis is almost identical to that of the BMD-1, the two-man turret has been replaced by a one-man turret with the gunner being seated on the left and provided with a one-piece circular hatch opening to the front. The reduction of the three-man crew to two means the vehicle can carry five passengers instead of four.

The main armament consists of a 30mm 2A42 dual-feed stabilised cannon with a 7.62mm PKT machine gun mounted coaxially to the right. This is the same weapon as that in the BMP-2. The BMD-2 carries 300 rounds of 30mm and 2,980 rounds of 7.62mm ammunition.

On the right side of the turret is a pintle which can take the AT-4 'Spigot' anti-tank missile, with a range of 2,000m, or it can take the AT-5 'Spandrel' which has a range of up to 4,000m. The BMD-2 only has a single 7.62mm PKT bow-mounted machine gun which is on the right side, while the left weapon port has been removed.

BMD-3 ACV
Just before the Soviet Union collapsed a final model of BMD appeared. In 1990 the BMD-3 entered service with Russian airborne units followed by the naval infantry. This featured a brand-new chassis fitted with the BMP-2 turret. Like its predecessors it is built from all welded aluminium, protecting the crew only from small-arms fire and shell splinters. Also like the BMD-2, it is armed with the 30mm 2A42 Cannon.

2S9 120mm Self-propelled Howitzer
This vehicle really comes under self-propelled artillery but as it is air transportable, can be paradropped and is based in the BMD (BTR-D) it is included here with the rest of the BMD family.

The 120mm SO-120 (2S9) *Anona* or Anemone self-propelled howitzer entered service in 1981 and was deployed in Afghanistan with Soviet forces fighting the Mujahideen. This was developed to meet the needs of the Soviet air assault divisions by providing an artillery and anti-tank capability. The *Anona* was armed with a turret-mounted 120mm 2A51 breech-loading mortar that had a 1.8m barrel. This had a fire rate of six to eight rounds a minute. Muzzle velocity was 560m/s for the HEAT round and 367m/s for the artillery rounds. Ammunition was fixed and loading was done manually, although the ramming was automatic. Once the round was in the feed tray, an electric button was pressed and a rammer automatically seated the round in the chamber and closed the breech.

The 2S9 took just 30 seconds to come into combat and a similar time to come out of action. When deployed to a firing position, the suspension was raised to provide a more stable firing platform. The fighting compartment had stowage for twenty-five rounds of ammunition. Mounted below the rear of the turret was an ammunition loading hatch and mounted on top of the hatch was a device for

A 120mm 2S9 self-propelled howitzer. This utilises the chassis of the lengthened BMD M1979, and both saw combat in Afghanistan during the 1980s.

loading ammunition from the ground directly into the fighting compartment. This permits continuous fire without depleting the onboard ammunition supply.

Its aluminium hull was a version of the BTR-D airborne ACV. The two-man turret was located above the fighting compartment, and was of welded aluminium construction with 16mm frontal armour. The turret roof had two hatches, one for the gunner and one for the loader; traverse was limited to 35 degrees either side. This vehicle lacked defensive weapons as it had no coaxial machine gun and there was no bow or turret-mounted machine gun.

The tracks were the same as those used on the BMD-1 and the suspension was hydraulic with adjustable ground clearance of 100mm and 450mm. As in the BTR-D there were six road wheels each side and five track return rollers. Although never exported, a number of 2S9s were bequeathed to the Afghan Army when the Soviets withdrew in 1989.

Chapter Eight

BRDM Amphibious Scout Car

In the late 1950s the Soviet Army's standard reconnaissance vehicles consisted of the Second World War-vintage BA-64 armoured car and the BTR-40, which had appeared in 1948. Neither was amphibious and they lacked adequate cross-country capabilities.

BRDM-1

The BRDM-1 amphibious scout car entered service in the late 1950s. Notably, the BRDM Model 1957 was open-topped while the Model 1958 was enclosed with twin roof hatches and the latter became the standard production model. The BRDM-1's hull was of all-welded steel with the engine located at the front and the crew compartment at the rear. The driver sat on the left-hand side at the front with the commander on his right. Both had a hatch that swung forward and had a vision block for use when closed. There were also vision slits in the front of the hull.

A BRDM-1 Model 1958 amphibious scout car. The Soviet Union built about 10,000 of these vehicles which included command and NBC variants.

This shows the pintle-mounted 7.62mm SGMB machine gun on the forward part of the roof of the BRDM-1. This mounting left the gunner exposed. Some vehicles were fitted with a 12.7mm heavy machine gun in this position, with the 7.62mm mounted to the rear.

There were two firing ports either side of the hull and two large hatches in the forward part of the roof that opened rearwards. The rear of the crew compartment sloped at an angle of about 30 degrees and was fitted with a two-piece hatch that opened either side of the superstructure. There was also a firing port in each hatch. The vehicle was normally armed with a 7.62mm SGMB machine gun, pintle-mounted on the forward part of the roof. Some were also fitted with a 12.7mm DShKM heavy machine gun mounted in the same place, with a 7.62mm machine gun mounted at the rear.

The vehicle had two belly wheels, located between the front and rear road wheels, that could be lowered to greatly improve its cross-country performance. Like most Soviet AFVs, the BRDM-1 was fully amphibious, being propelled through the water by a single water jet to the rear of the hull. A trim vane was stowed folded under the nose of the vehicle when not in use. The BRDM-1 served with about a dozen countries, including Afghanistan, Cuba, Mozambique and Sudan.

BRDM-1 Anti-Tank Guided Weapons

The BRDM was modified as an anti-tank guided missile carrier and appeared in two versions. The first successful Soviet anti-tank missile was codenamed AT-1 'Snapper' by NATO, but was known by the Warsaw Pact countries as the 3m6 *Schmel* or Bumble Bee.

This wire-guided missile was first mounted on a modified GAZ-69 (later known as the UAZ-69) 2½-ton truck in which the quadruple launcher was completely unarmoured. On the BRDM-1 it was possible to provide a reasonable amount of protection for a triple mounting of the same weapon. The upper part of the rear hull of the BRDM was extended to take a retractable launcher for three 'Snapper' missiles that stood clear of the hull top when in action and could be completely covered by sideways-folding plates when not in use. One reload of three missiles was carried inside the vehicle.

The follow-on radio-guided missile, known to NATO as the AT-2 'Swatter', was likewise mounted on the BRDM-1. Although the 'Swatter' was only slightly smaller than the 'Snapper', it was possible to accommodate a retractable quadruple launcher on the BDRM-1, that was fully protected when not in use by folding side and rear plates.

The two belly wheels on this BRDM-1 have been lowered; these were designed to improve cross-country performance and were a feature retained with the BRDM-2.

The BRDM-2 appeared in the early 1960s and was fitted with the same one-man turret as the BTR-60PB APC. This scout car and its various variants were widely exported.

BRDM-2

The BRDM-2 appeared in the early 1960s. This vehicle featured the same turret as that installed on the Soviet BTR-60PB and Czech OT-64 SKOT-2a APCs. However, a wide variety of turretless variants were also produced as platforms for various Soviet anti-tank guided weapons. In the standard BRMD-2, armament comprised a 14.5mm KPVT machine gun with a 7.62mm PKT machine gun mounted coaxially to the right. A telescopic sight was mounted to the left of the main armament.

The driver sat in the front on the left-hand side with the commander to his right. They were provided with a bulletproof windscreen to their front which could be covered by two armoured shutters hinged at the top. When the shutters were closed the driver and commander were served by a series of periscopes around the front and sides of the vehicle. Entry was by two circular hatches immediately behind the commander and driver that opened vertically toward the centre of the vehicle.

Like its predecessor, a central tyre-pressure regulator system permitted the driver to alter the tyre pressures to suit the terrain. The driver was able to adjust individual tyres or all four while the vehicle was on the move. Also as with the BRDM-1, on each side of the vehicle, between the front and rear wheels, were two chain-driven belly wheels, that were lowered by the driver to improve cross-country performance and allow for ditch crossing. The suspension was formed by four semi-

This BRDM-2 was photographed during a parade in 1983. The trim vane for travelling in water is stowed under the nose of the hull.

A BRDM-2 from the rear with the water jet cover in the closed position. This could push the vehicle through the water at 10km/h.

elliptical springs with telescopic dual action shock-absorbers mounted two per axle. Steering was hydraulically assisted on the front wheels with the sealed brakes having air-assisted hydraulic actuators. The vehicle was fully amphibious. The BRDM-2 was exported to around fifty countries and saw extensive combat.

BRMD-2 Anti-Tank Guided Weapons & Surface-to-Air Missile Systems
Variants of the BRDM were produced carrying the AT-2 'Swatter', AT-3 'Sagger' and AT-5 'Spandrel' anti-tank missiles. The 'Sagger' variant comprised a BRDM-2 with its turret removed and fitted with an arm on top of which were mounted six AT-3s. When on the move they were stored within the hull, but in combat the arm was raised above the hull, complete with overhead armour protection. The missiles could

Polish troops on exercise with a BRDM-2. Note the driver and commander's armoured shutters are closed, leaving them reliant on the periscopes. The Polish Army at one time had 800 BRDMs.

Hiding amongst the trees is what appears to be a BRDM-2U command vehicle. This carried a generator to power the communications equipment.

The BRDM-2 was first seen publicly in 1966. This had a much better all-round performance than its predecessor and was supplied to around fifty countries.

This intelligence photograph of a BRDM-2 was undoubtedly taken by a defence attaché during a military parade. Note how the censor has blanked out the background to hide the vehicle's exact location. The only access to the vehicle is via the roof hatches just behind the commander and the driver. Just below the turret are three vision blocks that protrude from the outside of the hull, and in front of these is a single firing port.

either be launched from within the vehicle or from up to 80m away with the assistance of a separation sight. A total of eight missiles could be carried in reserve. The 'Swatter' variant had a quadruple launcher. A total of eight missiles were carried including three in the ready-to-launch position. The 'Spandrel' variant has been called the BRDM-3. This carries five missiles ready to launch above the turret with another ten in the hull.

The SA-9 Gaskin mobile surface-to-air missile system was based on a modified BRDM-2 chassis with its belly wheels removed. The original turret was replaced by a one-man turret with an elevating arm on either side on which were mounted two box-type launchers for the SA-9 fire-and-forget missiles.

BRDM-2s with five 'Spandrel' anti-tank missiles in the ready-to-launch position.

Chapter Nine

Self-Propelled Artillery

For a long time the Soviet Army clung to its belief in towed artillery. During the Second World War the American, British and German armies all developed self-propelled guns, but the Soviets chose to focus instead on assault guns and tank destroyers. Then in the mid-1970s it became apparent that the Soviets had changed their thinking when they produced two highly-capable weapons systems known as the 152mm M-1973 and 122mm M-1974 self-propelled guns. These were to see extensive combat around the world thanks to the superpower-backed regional conflicts fought in Africa and the Middle East.

122mm M-1974 (2S1) Self-propelled Howitzer

The 122mm 2S1 self-propelled howitzer, dubbed the M-1974 in the West, in fact entered service with the Soviet Army in 1971. Likewise the 152mm 2S3, called the M-1973, also entered service that year. The Soviet military designation for the 2S1 (its industrial number) was SO-122, although it was more commonly known as the *Gvozdika* or Carnation. As both these vehicles had turrets they were commonly mistaken for tanks. A way to tell them apart is that self-propelled guns normally have their turrets set at the rear of the hull rather than the middle, which is normally the case with tanks.

The M-1974 shared many of the same automotive parts as the MT-LB multipurpose tracked armoured vehicle. The suspension system was similar to the MT-LB and consisted of seven road wheels with the drive sprocket at the front and the idler at the rear; it had no track return rollers. Its all-welded steel hull was divided into three compartments; driver at the front on the left, engine behind the driver and the turret at the rear.

The main armament, designated the 2A31, was a modified version of the 122mm D-30 towed howitzer. This was fitted with a fume extractor and muzzle break and was held in position when travelling by a lock on the hull glacis plate, operated by remote control by the driver. The ammunition mix normally consisted of thirty-two HE, six smoke and two HEAT-FS rounds. It is believed around 10,000 2S1 were built

This 2S1 M-1974 belonged to the Iraqi Army and was abandoned during Desert Storm in 1991.

A Polish 2S1 on the firing range. This self-propelled howitzer was built by the Soviet Union, Bulgaria and Poland with production ending in the early 1990s.

for the home and export markets with production coming to an end in 1991. This vehicle was widely exported to Africa and the Middle East.

152mm M-1973 (2S3) Self-propelled Gun Howitzer

The 2S3's main armament consisted of a 152mm gun designated the 2A33, based on the 152mm D-20 towed gun howitzer, but with a bore evacuator fitted just behind the muzzle brake. Depending on the type of ammunition used, this had a maximum range of 24,000m. Just over thirty rounds, complete with fuses, were stored in the rear of the hull in three horizontal layers.

The large all-welded turret had a sloped front and sides with vision blocks on each side. The commander sat on the left of the turret and the loader on the right. The commander's cupola could be rotated 360 degrees and had a single hatch opening to the rear. Mounted on the forward part of the hatch was a 7.62mm PKT machine gun that could be aimed and fired from inside the turret. To the left of this was normally mounted an OU-3K infra-red/white light searchlight.

This 152mm 2S3 or M-1973 self-propelled howitzer was photographed with its Soviet crew in Afghanistan. It also saw combat with Iraq, Libya and Syria.

An Iraqi 152mm 2S3 M-1973 self-propelled howitzer also captured in 1991. Visible on the left is an MT-LB and to the right a BMP.

The torsion bar suspension consisted of six dual rubber-tyred road wheels on either side, with the drive sprocket at the front, the idler at the rear and four track return rollers. The first and last return rollers only supported the inside of the track. About 10,000 of these self-propelled guns were also produced by the time the Soviet Union folded. Notably, the 2S3 served with the Hungarian, Iraqi, Libyan and Syrian armies.

240mm M-1975 (SM-240) (2S4) Self-propelled Mortar

The turretless 2S4, 2S5 and 2S7 all looked quite similar, featuring a mortar mounted on the top of the hull, though the 2S7 was by far the largest. In particular the 2S4 and 2S5 used the same chassis and running gear. The Soviet 240mm self-propelled mortar was dubbed the M-1975 by the West, but was known as the SM-240 (2S4) by the

Soviet Army who tended to call it the *Tyul'pan* or Tulip Tree. The chassis for this was a modified version of that used in the SA-4 'Ganef' surface-to-air missile system.

The hull of the 2S4 was of all-welded armour construction that gave the crew protection from small arms fire and shell splinters. The 240mm smoothbore mortar was transported complete with its base plate on top of the hull in a horizontal position. The mortar could be hydraulically lifted by remote control to the rear of the vehicle so that in its firing position it faced rearward. Some forty mortar bombs were carried in two drum magazines which are off-loaded via a hatch in the roof. The mortar could fire conventional high explosive fragmentation bombs or an HE FRAG rocket-assisted projectile, which had a range of 18,000m. Only about 400 2S4s were built and some were supplied to the former Czechoslovakia, Iraq and Lebanon.

152mm 2S5 Self-propelled Gun

The 2S5 known, as the *Giatsint* or Hyacinth, went into service in 1978 and had the same suspension and running gear as the 2S3. The long-barrelled 152mm gun,

The 152mm 2S5 Hyacinth self-propelled gun in travelling configuration with the ordnance locked in position. It was deployed in batteries of six weapons, three batteries making up a battalion.

designated the 2A37 and fitted with a five-part multi-baffle muzzle brake, was mounted externally on the roof to the rear and had a maximum range of 37,000m.

The chassis was of standard all-welded steel construction with a maximum armour thickness of 13mm. The driver was seated at the front on the left and had a single-piece hatch opening to the rear. In front of this were periscopes. The commander sat in a raised superstructure to the rear of the driver and had a cupola that could traverse through 360 degrees. The remaining three crew members were seated in a compartment at the rear of the hull accessed via a ribbed ramp in the rear. This compartment was fitted with roof hatches and periscopes that give fields of view to the sides of the vehicle.

Ammunition was of the separate-loading type, i.e. projectile and propellant charge, and the crew were assisted by a semi-automatic loading system. This comprised an electrically-driven chain rammer to the left of the breech, which folded back through 90 degrees to be parallel with the breech. A charge-loading system was pivoted on the right side and had a projectile tray and charge tray. The ammunition was loaded into the trays and then swung upward through almost 90 degrees where the rammer first rammed in the projectile and then the charge. The 2S5 could carry thirty rounds, with the projectiles stowed vertically in a carousel to the left of the rear compartment and the charges to the right in three rows of ten. Finland was the only known export customer for this self-propelled gun.

203mm (SO-203) (2S7) Self-propelled Gun

This self-propelled gun came into service in 1975 with well over 1,000 being built. A fully-enclosed crew compartment was at the front with seats for the commander and the driver. They were each served by a circular hatch in front of which were periscopes for forward observation. They also had windscreens that could be covered by an armoured shutter hinged at the top when in action.

The 203mm gun was designated the 2A44 and had no fume extractor or muzzle brake. The gun operator sat at the rear on the left side with the elevation and traverse controls. He was served by a standard PG-1M panoramic telescope that was used with the K-1 collimator. For direct fire he employed the OP4M-87 telescope. The weapon had an ammunition handling system that permitted a rate of fire of two rounds per minute out to a range of 47,000m. At the rear of the vehicle was a very distinctive hydraulically-operated spade that is lowered before firing starts. The 2S7 took up to six minutes to deploy for action, which was not ideal.

The V-12 liquid-cooled diesel engine was at the back of the crew compartment. To the rear of the engine was a second crew compartment that took four other personnel, accessed by two circular roof hatches. The torsion bar suspension comprised seven dual rubber-tyred road wheels either side, with six track return

A very poor quality intelligence photo of the massive 203mm 2S7 self-propelled gun. This came into service in 1975 and about a thousand were built. Some were supplied to Czechoslovakia and Poland.

rollers with the idler at the rear and the drive sprocket at the front. This self-propelled gun was only deployed by the Soviet Army and a few other Warsaw Pact countries.

152mm (2S19) Self-propelled Artillery System

The 2S19 152mm self-propelled gun went into service just as the Soviet Union was disintegrating. Its chassis is based on the running gear and suspension from the T-80 and the power-pack from the T-72. The main armament consists of a turret-mounted long-barrelled 152mm gun, the 2A64, fitted with a flume extractor and muzzle brake. When in motion the barrel is held in position by a lock mounted on the front glacis plate. The turret is very large but can manage a 360-degree rotation.

Polish-built T-54/55s on exercise. The first tank to be built in Poland was the Russian designed T-34/85: this was followed by the T-54 and then the T-55, with 8,570 being built at Labedy. The bulk of these were exported. The Polish T-54/55s are readily distinguishable from their Russian cousins by the large rectangular stowage box on the left side of the turret, clearly visible on these examples.

Chapter Ten

Anti-Tank Missiles

From the late 1960s wire-guided anti-tank missiles became a feature of the Soviet and Warsaw Pact armies. These were carried on a variety of platforms and were designed to give their tank and motor rifle divisions greater punch. In particular Moscow wanted to ensure that in the event of a massive tank battle on the central German plain, its infantry would have the capability to take on NATO's armour. It was the AT-3 'Sagger' that first demonstrated the real threat posed by such weapons by knocking out numerous Israeli tanks during the 1973 Yom Kippur War.

AT-1 'Snapper'

The AT-1, dubbed the 'Snapper' by NATO, was the very first Soviet anti-tank guided weapons (ATGW). The missile had a single-charge solid motor and four large

The BRDM-1 2P27 was modified to carry the AT-1 'Snapper'.

Dubbed the AT-1 'Snapper' anti-tank missile by NATO, the Soviet designation was 3M6 *Shmel* or Bumblebee. This was a manual-command-to-line-of-sight guided missile.

cruciform wings. It had an effective range of 2,500m and its 5.25kg hollow-charge warhead could penetrate 380mm of normal steel, but it was wire-guided and had the disadvantage that the operator had to watch both the missile and the target throughout the missile flight. On firing, the missile flew like a fast model aircraft, paying out the fine wires behind it which were attached to the operator's joystick.

Three 'Snappers' were fitted to the turretless BRDM-1 amphibious scout car, mounted on launch rails over the rear of the hull (see Chapter 8). The AT-1 saw service with the Arab armies during the 1967 Six Day War and many were captured intact by the Israelis.

AT-2 'Swatter'

The radio-guided AT-2 'Swatter' appeared some years later and had four wings, also in cruciform but smaller than on the AT-1. All the wings were fitted with elevons or control surfaces, with two carrying tracking flares to assist with guidance during flight. A solid-fuel motor with oblique nozzles between the wings fired the missile off a large launch rail. The warhead had a very blunt nose with two small fin-like projections. The 'Swatter' had the same range as 'Snapper' but could penetrate 480mm of armour. The improved 'Swatter-C' had an extended range of 3,500m and was converted from its original radio-command-to-line-of-sight guidance to semi-active infra-red guidance.

The AT-2 equipped the BRDM-1 and BRDM-2 amphibious scout cars as well as the M-24 helicopter gunship. On the BRDM it was in a quadruple ready-to-launch configuration (see Chapter 8). Like the AT-1 the AT-2 saw extensive combat during the Arab-Israeli wars. It was also used on the Mi-24 in Angola, Afghanistan, during the Iran-Iraq War and the 1982 Lebanon War. Both the 'Snapper' and 'Swatter' were rendered obsolete by the much-improved AT-3.

AT-3 'Sagger'

The AT-3 'Sagger', called the *Malatyuka* by the Soviets, was first spotted in a Moscow parade in May 1965. The missile was launched by a boost motor just behind the warhead that had four oblique nozzles. It had no aerodynamic controls but the small wings could fold in for transport. The operator was guided by a tracking flare attached to the body and it could be steered visually out to 1,000m and to three times that with a magnifying optical sight.

The 'Sagger' was used to arm the BRDM with a six-round retractable launcher, the BMP and BMD with a single reloadable launcher above the main gun, and the Czech SKOT with a twin reloadable rear launcher. The Mi-24 Hind could also carry this missile on its four outboard launcher pylons. The 'Sagger'-armed BRDM-2 played a key anti-tank role in the Soviet Army. Each tank division had nine of them to support its BMP-equipped motorised infantry regiments. Likewise, each motor rifle

'Sagger' and launch rail installed on the main gun of a BMP-1.

A Romanian BRDM-2 firing 'Sagger', the top plate protects the launcher when retracted into the vehicle. The Romanian Army fielded about 400 BRDM-1/-2 in 1985.

The man-portable version of the AT-3 'Sagger', consisting of the missile, launch rail, joystick control unit, sighting mechanism and carrying case. This first made its mark in the Yom Kippur War of 1973.

division had thirty-six BRDM-2s armed with ATGWs, nine in the anti-tank battalion, nine in the BMP-equipped motor rifle regiment and nine in each of the two motor rifle regiments equipped with the BTR. The missile could also be fired from a simple ground launcher.

The 'Sagger' gave the Israeli Defense Force a nasty shock in 1973 when two-man teams of Egyptians operated it from a small portable launch platform. Each Egyptian infantry division included an ATGW battalion equipped with 48 'Saggers' and 314 RPG-7 anti-tank grenade launchers. An individual soldier could carry the 'Sagger' missile and the case converted into a launching platform connected to the joystick. The main problem controlling the 'Sagger' and other wire-guided ATGWs is in gathering it on to the line of the target after launching from a remote position. This takes about 400m, and once on target the operator needed nerve as much as skill to keep it there.

AT-4 'Spigot'

Codenamed 'Spigot' by NATO, the AT-4 was a high-performance tube-launched missile similar to the Euro Milan missile that first appeared in 1970. The design of the launcher ensured the operator could remain under cover while firing the missile: once launched, only the launcher's black tracking head remains visible. The missiles were carried in their launch tubes, which were discarded after firing.

The system was widely deployed by the Soviet and other Warsaw Pact armies. In the man-portable configuration it weighed around 40 kilos. Control was wire-guided semi-automatic command to line-of-sight (SACLOS), where all the gunner has to do is keep the cross-hairs of his sight on the target to ensure a hit, rather than actively fly the missile onto the target as with first-generation ATGWs. Range was estimated to about 2,500m. The AT-4 was later supplemented by the AT-7 'Saxhorn' which was lighter and easier to deploy.

The 9M111 *Fagot*, better known by its NATO name as the AT-4 'Spigot', entered service in 1970. This particular example is being deployed by the Polish Army, which was equipped with the 'Snapper', 'Sagger' and 'Spigot'.

A Romanian BRDM-2 launching AT-5 'Spandrel' missiles.

AT-5 'Spandrel'

The AT-5 'Spandrel' ATGW was a second-generation SACLOS type and was first seen during a Red Square Parade on 7 November 1977. It was a tube-launched system and was mounted on the BRDM-2 with five ready-to-launch missiles. The tube was similar to that of the Milan, with a blow-out cap at the front and a flared tail through which passes the efflux from the boost charge. This blows the missile out of the tube before the ignition of its own flight motor. The BRDM carried ten reloads. The AT-5's general similarity to Milan was probably not a coincidence.

The HEAT warhead weighed around 7kg and could penetrate up to 600mm of armour. This wire-guided missile homed in on its target by an infra-red heat-seeking system. A ground mount was always carried to allow 'Spandrels' to be launched away from a vehicle. It had a maximum range of 4,000m. By the late 1970s the Group of Soviet Forces in Germany had significantly enhanced its anti-tank capability by replacing all its 'Swatter' and 'Sagger' missiles with 'Spandrels'.

Some BRDM-2s carried a mix of ATGWs. For example, some Iraqi Army BRDM-2s were armed with three AT-5 'Spandrels' on the right and two AT-4 'Spigots' on the left. This enabled the vehicle to carry a larger missile payload: rather than ten AT-5s it could accommodate eight AT-4s and six AT-5s. Exported BMP-2s were supplied with the 2,500m AT-4 'Spigot' ATGW instead of the 'Spandrel' installed on Soviet vehicles. The 'Spigot' version was distinguishable from the 'Spandrel' by its shorter launch tube, which was straight, while the Spandrel was about 200mm shorter with a tapered end.

The Soviet Union revealed in 1990 that it had deployed a version of the MT-LB multipurpose armoured vehicle armed with the AT-6 'Spiral'. The vehicle has a gunner's sight mounted at the front of the hull and a single retractable rail launcher at the rear.

AT-6 'Spiral'

The 'Spiral' entered service in 1976 and was designed for use with the Mi-24 'Hind' helicopter gunship (see Chapter 11). It was subsequently installed on the MT-LB and there was also a shipborne version. Delays in the development of this missile initially meant that the Mi-24 had to be armed with an improved version of the AT-2. The SACLOS missile had a small booster stage to launch it from the tube and then deployed a solid-fuel sustainer rocket. It had a radio command link which gave greater speed and range than the more traditional wire guidance. The launch range was 400m out to 5,000m and the HEAT warhead could cut through over 550mm of armour. Understandably this made it a missile greatly to be feared.

AT-7 'Saxhorn'

The AT-7 'Saxhorn' (9k115 *Metis* or Mongrel) man-portable SACLOS wire-guided anti-tank missile went into service with the Soviet Army in 1979 to supplement the AT-4. The 'Saxhorn' is much lighter than the latter, with a simpler tripod launcher, but only has half the range. This missile system, which includes the 9P151 launching post, was issued to Soviet motor rifle companies. NATO dubbed the improved version of the AT-7 which appeared in the 1980s the AT-13 'Saxhorn-2'.

This MT-LB is launching the AT-6, known to the Soviets as the 9k114 *Shturm* or Assault.

The AT-7 'Saxhorn' (9k115 *Metis* or Mongrel) went into service with the Soviet Army in 1979 to supplement the AT-4.

AT-8 'Songster'

When experiments with purely missile-armed tanks ended in failure, the Soviets looked to a hybrid system. The result was a SACLOS anti-tank missile that could be fired from the 125mm gun of the T-64 and T-80 tanks. Officially known as the 9K112 *Kobra* by the Soviets it was first installed in the T-64B in 1976 and the T-80B two years later. The T-64B, which could carrying six AT-8s plus thirty-six gun rounds, could fire up to four 'Songsters' a minute. The T-80B could only carry four missiles, but the T-80U was able to carry a similar number to the T-64.

The missile came as two separate sections and was stored in the autoloader in the same way as the conventional 125mm rounds. It could engage ATGW systems, tanks and even helicopters. The radio transmitter for the AT-8 was mounted in a steel box in front of the right commander's cupola, though the missile was guided to the target by the gunner. It was propelled out of the barrel by a boost motor; the main motor then cut in and powered it to the target. 'Songster' had a muzzle velocity of 125m/s that increased up to 800m/s and took just ten seconds to cover 4,000m to penetrate 600mm of steel armour. It was not assessed to be very effective against ceramic and reactive armours, however.

The 'Saxhorn' weighs less than the 'Spigot' by virtue of having a lighter missile and more simple tripod launcher.

Chapter Eleven

Anti-Tank Helicopters

The Soviets were slow to appreciate the merits of attack helicopters. This may have been in part because their helicopter forces came under the control of the air force, which had no desire to develop a helicopter that duplicated the role of its strike aircraft. Nonetheless, they were soon to catch up and produced one of the most iconic helicopter gunships of the Cold War – the Mi-24. Due to its very distinctive stepped front it was nicknamed the *Gorbach* or hunchback by its crews. It was to become a common sight in many of the world's regional conflicts and has seen extensive combat. Its considerable firepower ensured that it was a weapon system to be feared and respected in equal measure.

Mi-4/Mi-8 'Hip'

Initially, the Mi-4A helicopter developed in the early 1960s had a single machine gun mounted in a gondola underneath the fuselage and could carry bombs. The limited field of fire of the machine gun greatly restricted the Mi-4's utility as a gunship. Similarly,

The Mi-8TB 'Hip-E' helicopter showing the lateral pylons capable of taking three rocket pods either side with a total of 192 rockets. The armed 'C' variant was a popular export model and by the mid-1980s around 7,000 Mi-8s had been sold to over forty countries.

An armed 'Hip' engaging targets in Afghanistan – 'Hips' saw extensive combat against the Mujahideen in a variety of roles along with the Mi-24 gunship.

The business end of the 'Hip-E' gunship showing the nose machine gun, the six hardpoints for a mix of ordnance that includes six rocket pods, plus the launch rails for four AT-2 'Swatter' ATGMs.

attempts at fitting the Mi-1 utility helicopter and the Mi-4 transport helicopter with anti-tank guided missiles and rockets proved largely unsuccessful.

Introduced in the late 1960s, there were numerous variants of the Mi-8 utility helicopter, dubbed the 'Hip' by NATO. The 'Hip-A' and 'B' were the prototype and basic models, while the 'C' was the first mass-produced utility transport version. The latter could carry four unguided rocket pods. The gunship version was known as the 'Hip-E' and at the time was called the world's most heavily-armed helicopter. Its weaponry included a 12.7mm machine gun in the nose, a triple stores rack on each side capable of carrying six rocket pods, whilst simultaneously carrying four AT-2 'Swatter' anti-tank missiles. The export variant was known as the 'Hip-F' with the armament changed to six AT-3 'Sagger' missiles.

Experiments were also conducted with the Soviet Mi-2 built by WSK-Swidnik in Poland in the early 1970s. The Mi-2 attack helicopter was armed with four 'Sagger' missiles mounted on pylon launchers, a 7.62mm PKM machine gun mounted in the side cargo door and a fixed forward-firing PKM in the lower port fuselage.

Soviet developments were mainly in reaction to what the West was doing. The M-4 transport helicopter came about because of American experiences with helicopters during the Korean War, and it was the use of US attack helicopters in Vietnam that finally caught the Soviet Union's attention. In the summer of 1968 the Helicopter Directorate of the Soviet Army's General Staff issued a technical

The 'Hip', like the 'Hind', survived the Cold War and remains in service to this day. This one was supporting Coalition forces in Afghanistan.

requirement for the development of an attack helicopter to the Mil design bureau.

This requirement was given some impetus when Soviet KGB Border Guards found themselves fighting the Chinese Army along the Ussuri River the following year. The worry was that the Chinese might attempt to cut the strategic Trans-Siberian Magistral Railway if the border clashes escalated into all out war. Understandably, the KGB felt that a heavily-armed helicopter would be ideal for patrolling the enormous Soviet-Chinese frontier.

Mi-24 'Hind'

In 1970 the Mi-24 prototypes took to the air for the first time. This essentially was designed as a rocket platform, armed with a single 12.7mm machine gun and very thinly armoured. NATO dubbed this the 'Hind-B', but it was quickly followed in 1972 by a series production version called, rather confusingly, the 'Hind-A' by NATO. This had reconfigured sub-winglets, in part to give the helicopter a higher cruise speed and better lift, and a more powerful engine. It could also take the *Falanga* (AT-2 'Swatter') anti-tank guided missile.

It was first seen in large numbers in East Germany in 1974 with the Soviet Western Group of Forces. However, the 'Hind-A' proved far from ideal: the rotor

During the Cold War Mi-24 helicopter gunship variants were dubbed 'Hind-A' to 'Hind-E' by NATO. This is the transport version known as the 'Hind A' and was deployed with the Group of Soviet Forces in Germany.

A 'Hind-D' gunship anti-armour helicopter showing its UB-32 rocket pods as well as its four-barrelled heavy machine gun in the under-nose turret – the twin 'Swatter' rails do not have any missiles. Over fifty countries operated the 'Hind'.

blade was vulnerable to small-arms fire while the weak tail boom could fracture. Once more the armour was inadequate as machine-gun rounds could penetrate the sides and the 'Hind-A's' own gun had limited traverse.

But the Mil design bureau was quick to learn from American experiences. Notably. they were able to lay their hands on components from the American AH-I Cobra attack helicopter deployed in Vietnam. The M-24's nose was redesigned in a tandem configuration just like the Cobra and the AH-56 Cheyenne. For better crew protection, armoured glass was installed on the forward canopies and the crew positions were placed in an armoured 'bathtub', which protected them against small-arms fire up to 7.62mm. Likewise, at the front the firepower of the Hind was significantly enhanced with the installation of a 12.7mm multi-barrelled heavy machine gun in a traversing barbette, flanked by the missile radio command antenna and a new night sensor port. The new-model gun had a rate of fire of 3,200 rounds a minute. The four inboard weapon pylons could carry various payloads while the two outboard pylons could take anti-tank missiles.

This new Mi-24 was codenamed the 'Hind-D' by NATO and included improvements in infra-red suppression and ballistic protection of the power train subsystem. It was soon nicknamed the *Gorbach* by its crews and was first seen in the West in 1976 but most likely entered service significantly earlier. When the new

A sight that would have filled NATO tank crews with fear. The 'Hind D' and 'E' have a tandem cockpit layout for the pilot and the gunner. By the early 1980s around 1,500 had been produced with about 250 going to Warsaw Pact and Soviet client states.

AT-6 'Spiral' ATGW appeared, the Mi-24 was adapted for the guidance equipment for this missile as the 'Hind-E'. The Mi-24 was widely exported to the Warsaw Pact and Soviet client states. The 'Hind' went on to see extensive action, particularly in the Soviet-Afghan War, the Iran-Iraq War and the Lebanon War and more recently in the Libyan and Syrian uprisings.

Soviet air power played a fundamental role in Moscow's operations in Afghanistan and there can be little doubt the cutting edge of the Soviet presence was the helicopter. Gunships provided swift and devastating firepower with a mix of machine guns, bombs, missiles and rocket pods. The Mi-24 'Hinds' conducted many of the close air support operations, while the Mi-8 'Hips' supported by the larger M-6 'Hooks', conducted most of the troop-carrying and re-supply missions. The first reported use of Soviet anti-tank helicopters came in Afghanistan in 1979 when Mi-24s knocked out Afghan tanks around the presidential palace in Kabul. Within a very short time the Soviets had some 600 helicopters, including 200 Mi-24s, supporting operations in Afghanistan.

The Soviets did not have it all their own way, however, and had to quickly adopt nap-of-the-earth tactics, as Mi-24s were reportedly lost to SA-7 surface-to-air missiles as early as 1980. Pakistan even claimed to have shot down an Afghan 'Hind-A' straying over the border with a M55 quad 0.50 machinegun. Whilst only two dozen helicopters were lost by 1983, the growing use of heat-seeking missiles was the death-knell of the Soviet helicopter presence in Afghanistan.

In post-Soviet Afghanistan from 1992 onwards an unknown number of helicopters and fixed-wing aircraft were either shot down or destroyed during the civil war. When the Taliban stormed the HQ of Gulbuddin Hekmatyar's forces at Charasyab in 1995 they captured a number of helicopters. However, with the loss of Bagram air base in 1997 the Taliban Air Force was forced to destroy or disable many of its aircraft. Estimates for losses by all causes between 1992 and 1998 include about eighty transport helicopters and at least twelve Mi-24 gunships.

The 'Hind' has seen extensive combat around the world and examples such as this Mi-24P remain in service with the Russian Air Force. After the Cold War it was deployed in the troubled Transcaucasus region of the Russian Federation during the innumerable wars fought there.

A modern Russian Mi-24PN. Moscow planned to replace the Mi-24 with the Mi-28N and Ka-52 but there were long-running development problems.

The Afghan Air Force flying an export version known as the Mi-35. Following the Soviet-Afghan War the Mi-24 has continued to see combat in Afghanistan. The Afghans took possession of six refurbished Mi-35s from the Czech Republic in 2008.

This Libyan Mi-25 (an export variant) 'Hind-D' along with two others was captured at Quadi Doum airfield in March 1987 during the Chadian-Libyan conflict. The Libyan air force was equipped with the Mi-24A and Mi-25. Another Libyan 'Hind' was captured at the Maaten al-Sarra airbase on 5 September 1987, much to the annoyance of Moscow.

The Mi-24 'Hind', described somewhat extravagantly as a 'flying battlecruiser', continues to be extremely popular in Africa with the region's resource-strapped militaries. For example, the Sierra Leone Air Force was operating two Mi-24V 'Hind-Es' in 2001 against RUF rebels until one crashed. Likewise Burundi acquired three 'Hinds' which are thought to have come from Ukraine. Whilst during the Ethiopia-Eritrea conflict of 1998–2000 Ethiopia lost six Russian-supplied Mi-24s to Eritrean ground fire.

In the late 1990s Zimbabwe obtained ten Mi-24/35s, which operated from Kariba, Democratic Republic of Congo (DRC). However, Russian pilots and technical staff assigned to Thornhill Air Force Base were recalled after Zimbabwe failed to service its $35 million debt for the purchase of the helicopters. At least two are thought to have been lost in the DRC and only half of those remaining were serviceable by mid-2001 when the Russian technicians left.

Work on the Mi-28 anti-armour attack helicopter, known as the 'Havoc' by NATO, commenced in the 1980s but it lost out to the Ka-50.

An early intelligence photograph of the Ka-50 *Chornaya Akula* or Black Shark, this was designated the 'Hokum A' by NATO. It did not enter service until 1995, after the Soviet Union's collapse,

Although the Mi-28 'Havoc' attack helicopter undertook its first flight in 1982, it did not enter service until 2006.

Chapter Twelve

Soviet Equipment in Combat

Moscow's military aid to its allies throughout the Cold War came at a severe price. Ultimately it was a way in which the Soviets could exercise influence and create dependency amongst many developing-world countries that lacked the technical ability to support Soviet-supplied weapons systems. Wherever there were Soviet weapons, there were invariably Soviet advisors and technicians. It also meant that for decades the Soviet Union and the United States of America were able to fight a series of very bloody proxy wars without ever directly coming to blows themselves.

In Africa and Central America, where Soviet allies fought interminable bush wars against well-armed and organised guerrilla armies, the tank was largely useless. It took the Marxist governments of both Angola and Ethiopia a long time to wake up to the fact that fighter aircraft were in fact much more useful. Then again, it is easier to train tank crews than it is jet pilots. In the Middle East, massive Soviet tank supplies to Egypt and Syria made sense in light of the good tank country there and the fact that both were regularly embroiled in large conventional wars with Israel, which proved ideal testing grounds for new weapons systems. All these wars were characterised by significant tank losses, many being simply abandoned.

Libya deployed Soviet arms against Chad in 1983–8, losing $1 billion worth of weaponry in 1987 alone including 200 tanks and APCs as well as numerous tank transporters at Quadi Doum and another 100 armoured vehicles at Maartan-as-Sarra. Such losses seemed to matter very little. Afterwards, thanks to oil revenues, Libya acquired 2,000 T-55/62s as well as 1,000 BMPs and 750 BTRs.

Angola and Ethiopia were Moscow's most important regional allies and customers. In the late 1970s, when Ethiopia was at war with Somalia, the Soviets delivered 400 T-54/55s. Somalia, a Soviet ally, received 100 tanks before they fell out. In the case of Angola, in 1986 Soviet arms supplies were to the value of $2 billion and included tanks, IFVs and artillery. Two years later, when Cuban forces propping up the Marxist Angolan government were in the process of withdrawing, Cuba handed over T-54s to government forces. Likewise Mozambique, also blighted by civil war, spent $1 billion on weapons which included the PT-76 light tank.

Soviet-built armour proved ideal for the semi-literate and ill-trained armies of the Developing World. The acid test was how it faired in combat, although it was often let down by its crews. The Ethiopian Army captured this Somali T-54 tank during the Ogaden War.

During the 1970s Soviet and Cuban-supplied T-54s such as this fought with the MPLA guerrillas in Angola, first against the Portuguese colonial forces and then their rivals in UNITA.

Ethiopia, in the midst of a bitter and bloody civil war, also took receipt of T-55s and BTRs in the mid-1980s. By the end of the decade it had spent over $5.4 billion on weapons whilst its population starved. The delivery of equipment to the port of Assab for the 1985 campaign in Eritrea consisted of T-55s and APCs. Three years later, the writing was on the wall for Ethiopia's Marxist government after its forces suffered defeats at Afabet, Keren and Inda Selassie. At the latter two battles, rebels captured 100 tanks and 60 artillery pieces. By the late 1990s, Ethiopia was estimated to have in excess of 350 tanks, all of Soviet origin; Angola had 200 and Mozambique around 100.

This T-54/55 was knocked out in Ethiopia during the long and brutal civil war.

Cuba, Moscow's number-one ally in Uncle Sam's backyard, received arms worth billions of dollars. During the late 1950s Cuba was sent T-34s, T-54s, SU-100s and BTR-60s. Cuban troops saw action during the Angolan and Ethiopian civil wars. In 1983, shipments included T-62s and Cuba ended up with some 1,500 tanks that it had little use for, except for re-export or use in Angola. By the 1980s, much of Cuba's armour was ageing and Moscow sent limited supplies of T-62s, PT-76s, BTR-152s and BTR-60s as replacements.

Similarly, in 1983, Nicaragua's Marxist government took delivery of T-55s, BTR-50/60s, armoured cars and artillery pieces. The following year they took receipt of another 150 tanks, which were followed by a further 1,200 vehicles in a massive deal worth $600 million. However, it was Soviet Mi-24 helicopters that helped the government defeat the Contra rebels, not tanks.

In Asia, both India and Pakistan received T-54 tanks from Moscow in the 1960s. They amassed 1,420 and 820 tanks respectively, which they used against each other in 1971. By the 1990s the Indians had about 500 T-54/55s and Pakistan 50 T-55s plus 1,300 Chinese copies. India re-equipped its massive armed forces with T-72s and BMP-2s, which it subsequently manufactured under licence. By the mid-1990s, India had 1,100 T-72M1 in service. This is a modernised T-72M with an additional layer of 16mm armour on the glacis plate and combination armour on the turret. It is easily identifiable by two ribs on the upper glacis plate. One-time Soviet satellite, Ukraine, sold the Pakistanis 300 T-80UDs in the mid-1990s, though they were probably built with Russian assistance.

Communist Vietnam, at loggerheads with all its neighbours in the wake of the Vietnam War, was strengthened in the mid-1970s by billions of dollars' worth of Soviet military aid. This included 200 T-62s, 1,000 T-55s and 300 BTRs. T-54s were also delivered to neighbouring Cambodia and Laos where communist guerrilla

Another T-54 lurking in the Angolan bush. The rifling can be seen clearly in the barrel of the gun. Soviet armour was involved in all of Africa's post-colonial bush wars.

A BTR-60 supplied to the Nicaraguan Army fighting the US-backed Contras. Vast quantities of Soviet equipment went to Latin America, most notably Cuba.

forces eventually came to power. Vietnamese tanks rolled into Cambodia in 1978 and stayed for ten years.

After the Soviet withdrawal from Afghanistan, Moscow poured in arms including T-62s and BTR-70s to prop up its puppet government. In a show of force the Afghan Army pushed an armoured column of over 600 vehicles, including 200 tanks, into Kabul in the spring of 1989. The Soviet logistics system not only enabled Moscow to sustain a ten-year war in Afghanistan, but to also prop up the communist government for a few more years before the Mujahideen and then the Taliban took over. Afghanistan received almost 1,000 Soviet tanks, 550 BMPs and 300 BTRs over

A BMP-2 with the Soviet Army in Afghanistan in the 1980s. This IFV had been developed with mechanised warfare in mind, not counter-insurgency. The Mujahideen often got the better of Soviet armoured columns and were experts at ambushes.

Soviet T-62 tank crews fraternising with soldiers from the Afghan Army. In Afghanistan tanks tended to be used as mobile pillboxes or for convoy escort duties.

These BTR-70s were photographed during the Soviet withdrawal from Afghanistan in 1989. Shortly, after with its economy in tatters, the Soviet Union imploded, finally ending the Cold War.

A beaming T-62 tanker trying to convince those at home that all is well – tanks were far from ideal for the fast-moving guerrilla war fought in Afghanistan.

the years – most of them lie smashed and scattered the length and breadth of the country.

Moscow also used middlemen such as Cuba, Syria, Libya and North Korea to deliver its weapons. The latter began licence production of the T-62 in the late 1970s that ran until the mid-1980s. Notably some items were not interchangeable with the Soviet-built vehicle. North Korea is also believed to have built the BMP-1. During the Iran-Iraq War, North Korea was encouraged to supply Iran with T-62, even though Iraq had long been a Soviet ally. Similarly many Eastern European countries exported vast quantities of licence-built Soviet armour. Once Iran invaded Iraq, Moscow ended its neutrality and began to supply Iraq again. British intelligence officers based in Kuwait watched as hundreds of tanks and APCs flowed north to fuel the war. Iraqi T-72s fared well against Iran's M48s, M60A1s and Chieftains. Reportedly the Iraqis only lost sixty T-72s during the entire war.

Cynically, the Soviets armed both North and South Yemen. In 1968 North Yemen received refurbished T-54s and South Yemen T-34s. North Yemen purchased $14

During the invasion of Iran in 1980 the Iraqi Army deployed the T-54/55 as part of its attacking armoured forces.

A BMP-1 photographed on the streets during the breakup of the Soviet Union. Soviet armour was embroiled in numerous wars of independence and civil wars as the Soviet republics went their own way..

Refurbished Iraqi BMP-1s following the war in 2003. Iraqi BMPs saw combat during the Iran-Iraq War and both Gulf Wars.

billion worth of weapons including T-62s in the mid-1980s. The two countries became a single republic in 1990 and total deliveries amounted to over 800 tanks, 270 BMPs and 300 BTRs.

Communist China was never able to compete on anything like the scale of the Soviet Union's armoured vehicle exports. Its derivative of the T-54, the Type 59, went into production in the late 1950s. During the Vietnam War the North Vietnamese were supplied with 700 Type 59s, Type 62 light tanks and Type 63 amphibious tanks.

Chinese Type 59s delivered to Pakistan had a good finish, but the tanks lacked gun stabilisation and power traverse, which impacted on the rate of fire. They also had relatively poor armour and manoeuvrability. Pakistan is believed to have upgraded some of these tanks with a 105mm gun. Other Chinese customers included Albania, Cambodia, Iran, Iraq and North Korea. Pakistan subsequently obtained 400 Type-69/85s. The Chinese also exported their follow-on Type 69 to Iran and Iraq. Despite producing the improved Type 85 and 90, China purchased 200 T-80Us from Russia in the early 1990s.

After the Second World War Czechoslovakia was permitted to build the T-34/85 tank and SU-100 assault gun largely for export to the Middle East. The T-55 was then licence built from the mid-1960s for domestic and export purposes, followed by the T-72 in the late 1970s. Czechoslovakia also exported upwards of 3,000 of its OT-64 APC, Iraq being the biggest customer. Similarly Poland built the T-34/85, T-54/55, T-62 and T-72. Additionally Bulgaria and Poland manufactured the Soviet MT-LB multi-purpose APC, numbers of which were exported to countries such as Iraq. Romania manufactured T-55s from the late 1970s.

Following its involvement in the Israeli-Lebanon war of 1982 Syrian President Hafez al-Assad claimed the T-72 was 'the best tank in the world'. However, it has been claimed that the Israeli Merkava and Syrian T-72s never came into contact and that most T-72 losses in Lebanon were due to ambushes. When the Israelis invaded

With some 70,000 built, the T-54/55 tank is the most iconic tank of the Cold War.

southern Lebanon in 1982 they found the Palestinian Liberation Organisation's armoury included sixty old T-34/85s and twenty T-54/55s. The Syrian Army caught up in the fighting lost 200 T-62s, 125 T-54/55s, nine T-72s and 140 APCs. At the time Moscow was highly displeased that the very first T-72 had fallen into Western hands. Afterwards, Syria went on a massive armoured vehicle buying spree. In the early 1980s Syria ordered 800 BMP-1s as well as large numbers of T-72s and BTR-80s. Then in the early 1990s the Syrians ordered 252 T-72s tanks from Czechoslovakia and another 350 from Russia. Syria ended up almost 5,000 T-54/62/72s.

Much upgraded, the T-72 has greatly extended its service life way beyond the Cold War. Iraqi T-72s subsequent tangled with the American M1A1 during the 1991 Gulf War and the 2003 Iraq War. On 26 February 1991, Iraqi T-72s gave the Americans a bloody nose during the Battle of Phase Line Bullet when they destroyed four Bradley IFVs and damaged another ten at a cost of six T-72s and eighteen APCs. Iran obtained a number of T-72s from Libya, North Korea, Poland and Russia during the 1990s and ended up with around 500. Likewise, India ended up with several thousand as insurance against another ground war with Pakistan. The T-72 also saw action during the Sri Lankan civil war, the Nagorno-Karabakh war, the civil war in Tajikistan, the various wars in former Yugoslavia and during the civil wars in Libya and Syria.

Soviet-supplied armour destroyed during the Arab-Israeli wars: the lower shot shows an Egyptian T-54 knocked out in 1973. Czechoslavakia and Poland built large numbers of T-54/55 tanks for the Arab armies in the Middle East.

Epilogue:
Money Not Tanks

It is hard to describe the Soviet threat during the 1970s and 1980s without it sounding simply like Cold War rhetoric. Nonetheless, regardless of Moscow's enormous nuclear weapons arsenal, Soviet manpower was sobering to say the least. By the late 1980s the Soviet Union had over four million men under arms, of which one and a half million were in the ground forces consisting of over fifty tank divisions, 150 motor rifle divisions and seven airborne divisions. Of these forces, some thirty divisions were deployed with the four Soviet groups of forces stationed in Eastern Europe; 200 to 300 tanks supported each division.

Key amongst these was the Soviet Western Group of Forces stationed in the German Democratic Republic and poised to strike the very heart of NATO. Its ground component comprised eleven tank and eight motor rifle divisions equipped with many of the AFVs described in this book. The air component consisted of twenty fighter and ground attack aircraft regiments.

Throughout the 1980s Washington produced an annual declassified report called *Soviet Military Power*, which was designed to highlight to decision-makers and the public alike the extent of the Soviet threat. In hindsight this was part of a concerted propaganda war designed to ensure Capitol Hill kept increasing US defence spending year on year. At the time though, *Soviet Military Power* stood as a very graphic testimony to the extent of the Red menace. Ironically, the Soviet Union was to be worn down by the economics of military confrontation with the West. Moscow's withdrawal from Afghanistan showed that in many respects the Soviet military was a bankrupt paper tiger. Within a couple of years, the Soviet Union imploded and was to re-emerge as the much-reduced Russian Federation.

It was President Mikhail Gorbachev who, by the late 1980s, realised that the Soviet Union could not win the arms race. When he tried to reign in defence spending, there was an attempted coup in 1991. The net result was that Russia, Belarus and Ukraine dissolved the Soviet Union and created the short-lived Commonwealth of Independent States. In the end Ukraine and all the Central Asian Soviet republics went their own way, leaving the Russian Federation. It was at this point the true scale of Soviet defence spending on its ground forces became apparent. Even after the dissolution of the Soviet Union, the Russia Army still had 55,000 tanks, 70,000 APCs, 24,000 IFVs and 9,000 self-propelled guns.

Moscow's policy of global arms transfers did not gain it any long-term influence during the Cold War. The only major leverage at the disposal of the Soviet Union was to cut off its arms deliveries, but in the case of Somalia this proved ineffective. Somalia went against Moscow's wishes over the Ogaden, as did Ethiopia over its handling of Eritrea. Furthermore, in the ever-growing international arms market, there were numerous alternative sources – not least Warsaw Pact members prepared to undercut Moscow. Moscow's allies were always aware of its limited economic clout. Lieutenant General Obasanjo of Nigeria in 1978 astutely noted, 'We need in Africa massive economic assistance to make up for the lost ground of the colonial era and not military hardware for self-destruction and sterile ideological slogans which have no relevance to our African society.'

Ironically, the Soviet Union's huge arms exports did not give Moscow any long-term strategic power. Egypt defected to the American camp, Libya and Syria became dangerous liabilities and while Libya moved to rehabilitate itself, Iraq and Syria fell into chaos. In Africa none of the Marxist governments in Angola, Mozambique and Ethiopia were able to militarily defeat the rebels, leaving their economies in tatters. Soviet support for Vietnam soured relations with China, while support for Cuba and Nicaragua aggravated America and contributed to Washington's decision to prop up unsavoury rightwing Latin American dictators.

Ultimately it was economic not military assistance that many of Moscow's clients really needed. This was something it was unable to provide and the gifting of billions of dollars worth of weaponry contributed to the collapse of the Soviet Union and bankrupted its arms industry. The Soviet war in Afghanistan proved to be a very public defeat for Moscow and highlighted how moribund the Soviet system was. Ultimately there can be no denying that the T-55 and T-62 became icons of the many regional conflicts fought during the Cold War – but at what price?

Suggested Further Reading

Beckett, Ian (general editor), *Communist Military Machine*, Twickenham, Hamlyn Publishing, 1985.

Bonds, Ray (ed), *Weapons of the Modern Soviet Ground Forces*, London, Salamander Books, 1981.

Buszynski, Leszek, *Soviet Foreign Policy and South East Asia*, Beckenham, Croom Helm, 1986.

Heritage, Andrew, *The Cold War: An Illustrated History*, Sparkford, Haynes Publishing, 2010.

Isby, David C., *Ten Million Bayonets, Inside the Armies of the Soviet Union*, London, Arms and Armour Press, 1988.

Miller, David, *Modern Tanks & Fighting Vehicles*, London, Salamander Books, 1992.

Moynahan, Brian, *The Claws of the Bear, A History of the Soviet Armed Forces from 1917 to the Present*, London, Hutchinson, 1989.

Urban, Mark L., *Soviet Land Power*, London, Ian Allan Ltd, 1985.

US Department of Defense, *Soviet Military Power*, Washington DC, Prospects for Change, 1989.

Suvorov, Viktor, *Inside the Soviet Army*, London, Hamish Hamilton, 1982.

White, B. T., *Wheeled Armoured Fighting Vehicles in Service*, Poole, Blandford Press, 1983.

White, B. T., *Tanks and other Tracked Vehicles in Service*, Poole, Blandford Press, 1978.

Zaloga Steven J., and George J. Balin, *Anti-tank Helicopters*, Oxford, Osprey Publishing, 1986.